About this book

This workbook contains practice to support your learning in P2/P3 maths.

Questions split into three levels of increasing difficulty – Challenge 1, Challenge 2 and Challenge 3 – to aid progress.

Symbol to highlight questions that test problem-solving skills.

Total marks boxes for each challenge and topic.

'How am I doing?' checks for self-evaluation.

Number – Number and Place Value

Counting in Steps of 2, 3, 5 and 10

Challenge 1

PS Problem-solving questions

PS **1** Answer these problems.

a) Each pencil case holds 10 pencils.

Count the total amount of pencils in lots of 10.
There are _____ pencils in total.

b) How many lots of 10 are there in 40? _____ lots

c) Complete the grid. Make sure you are counting in steps of 10!

	40			70			100

d) Start at 60 and count forwards 3 steps of 10. Which number are you at? _____

e) If 1 person has 10 toes how many toes do 3 people have? _____ toes

f) How many toes would 5 people have? _____ toes

g) How many people would 40 toes belong to? _____ people

Marks _____ /7

Challenge 2

1 Put the numbers in order to make a sequence counting **forwards** in steps of 10.

45, 75, 35, 15, 5, 65, 55, 25

2 Put the numbers in order to make a sequence counting **backwards** in steps of 10.

20, 50, 70, 60, 10, 30, 80, 40

PS **3** Each ladybird has three spots. The spots are counted in steps of 3.

a) Continue to count in steps of 3.

3, 6, 9, 12 _____, _____, _____

b) Now try this.

24, 21, 18, 15 _____, _____, _____

c) How many ladybirds would 6 spots belong to?

_____ ladybirds

Marks _____ /5

Challenge 3

PS **1** Complete this number grid. Check the steps each line counts in. Check whether they count forward or backward!

6		12		18
21		51		
	16		20	
67		61		55

Marks _____ /4

Total marks _____ /16   How am I doing? 😊 😐 😞

18   19

Starter test recaps skills covered in P2/P3.

Four progress tests throughout the book, allowing children to revisit the topics and test how well they have remembered the information.

Progress charts to record results and identify which areas need further revision and practice.

Starter Test

PS Problem-solving questions

1. Fill in the missing numbers. Check which way each set counts and the steps it counts in.

a)
3		5	6				10		12

b)
20		18		16			13		11

c)
2		6	8	10	12				

2. Put these numbers in order from lowest to highest.

a) 31 12 16 87 6

lowest   highest

b) 15 2 60 30 0

lowest   highest

c) 14 100 23 99 15

lowest   highest

3. Colour half of each shape.

a)   b)   c)

4

Progress Test 1

PS Problem-solving questions

1. Complete this addition grid. Make sure that each column and row has a total of 12. You can use the same number more than once.

	1	
	3	
	8	

2. Make ten different two-digit numbers using these single digit numbers.

Example: Digits 1 and 3 make 13

1 4 9 6 3

PS 3. Answer these number problems.

a) Savi has 6 caterpillars. He collects 7 more. How many does Savi have now? _____ caterpillars

b) The 7 caterpillars escape and Savi is back to 6. His friend Jemma takes three of them. How many caterpillars remain? _____ caterpillars

c) Each of Jemma's 3 caterpillars finds 5 friends. How many caterpillars does she have in total?

_____ caterpillars

34

Progress Test Charts

Progress Test 1

Q	Topic	✓ or ✗	See page
1	Numbers and Counting		13
2	Using Two-Digit Numbers		29
3	Solving Number Problems		39
4	Doubling and Halving using Addition and Subtraction		24
5	Numbers and Counting		13
6	Numbers and Counting		13
7	Place Value		22
8	Counting More or Less		20
9	Numbers and Counting		13
10	Doubling and Halving using Addition and Subtraction		24
11	Doubling and Halving using Addition and Subtraction		24
12	Using Two-Digit Numbers		29
13	Doubling and Halving using Addition and Subtraction		24
14	Counting in Steps of 2, 3, 5 and 10		18
15	Solving Number Problems		39
16	Counting in Steps of 2, 3, 5 and 10		18

Progress Test 2

Q	Topic	✓ or ✗	See page
1	What is Multiplication?		38
2	Place Value		22
3	Finding Fractions of Larger Groups		60
4	Doubling and Halving using Multiplication and Division		46
5	Doubling and Halving using Multiplication and Division		46
6	What is Multiplication?		38
7	Solving Multiplication and Division Problems		50
8	Counting in Steps of 2, 3, 5 and 10		18
9	Counting More or Less		20
10	Less Than, Greater Than and Equal To		24
11	Solving Number Problems		39
12	Finding Fractions of Larger Groups		60

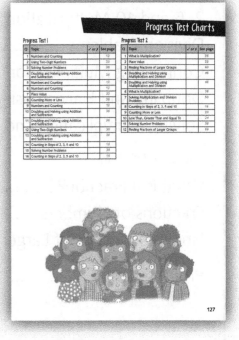

127

Answers for all the questions are included in a pull-out answer section at the back of the book.

Contents

Starter Test 4

Number – Number and Place Value

Numbers and Counting 12

Counting Forwards and Backwards 14

Counting in Steps of 2, 3, 5 and 10 16

Counting More and Less 20

Place Value 22

Less Than, Greater Than and Equal To 24

Number – Addition and Subtraction

Doubling and Halving using Addition and Subtraction 26

Solving Number Problems 28

Using Two-Digit Numbers 30

Solving Missing Number Problems 32

Progress Test 1 34

Number – Multiplication and Division

What is Multiplication? 38

What is Division? 40

2, 5 and 10 Multiplication Tables 42

Division Problems 44

Connecting Multiplication and Division 46

Doubling and Halving using Multiplication and Division 48

Solving Multiplication and Division Problems 50

Number – Fractions

What is a Fraction? 54

Fractions of Numbers 58

Finding Fractions of Larger Groups 60

Progress Test 2 62

Contents

Measurement

Measuring Length and Height.. 66
Measuring Weight and Volume.. 68
Comparing Measurements... 70
Measuring Temperature 72
Measuring Time... 74
Standard Units of Money 76
Money Problems... 78

Geometry – Properties of Shapes

2-D Shapes 80
3-D Shapes 82
Different Shapes 84

Progress Test 3... 86

Geometry – Position and Direction

Patterns 90
Sequences... 92
Quarter Turns and Half Turns... 94
Right-Angle Turns 96

Statistics

Pictograms.. 98
Tally Charts... 100
Block Diagrams... 102
Tables 104
Gathering Information and Using Data................................ 106

Progress Test 4... 108

Answers and Progress Test Charts (pull-out)...................... 113

ACKNOWLEDGEMENTS

The author and publisher are grateful to the copyright holders for permission to use quoted materials and images.

All illustrations and images are © Shutterstock.com and © HarperCollinsPublishers

Every effort has been made to trace copyright holders and obtain their permission for the use of copyright material. The author and publisher will gladly receive information enabling them to rectify any error or omission in subsequent editions. All facts are correct at time of going to press.

Published by Leckie
An imprint of HarperCollinsPublishers
Westerhill Road, Glasgow, G64 2QT

HarperCollins Publishers
Macken House, 39/40 Mayor Street Upper, Dublin 1,
D01 C9W8, Ireland

© Leckie 2023

ISBN 9780008665876

First published 2017

10 9 8 7 6 5 4 3 2 1

British Library Cataloguing in Publication Data.

A CIP record of this book is available from the British Library.

Printed in the United Kingdom.

MIX
Paper
FSC™ C007454

1. Fill in the missing numbers. Check which way each set counts and the steps it counts in.

a)
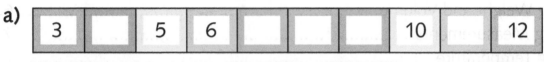

| 3 | | 5 | 6 | | | | 10 | | 12 |

b)

| 20 | | 18 | | 16 | | | 13 | | 11 |

c)
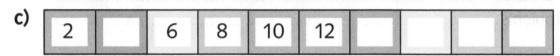

| 2 | | 6 | 8 | 10 | 12 | | | | |

3 marks

2. Put these numbers in order from lowest to highest.

a)

31   12   16   87   6

| | | | | |
lowest   highest

b)
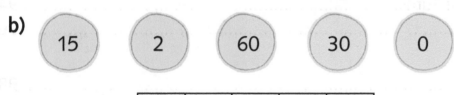

15   2   60   30   0

| | | | |
lowest   highest

c)
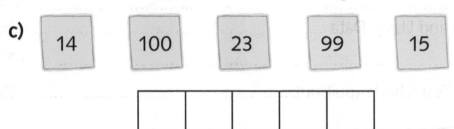

14   100   23   99   15

| | | | | |
lowest   highest

3 marks

3. Colour half of each shape.

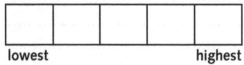

a)   b)   c)

3 marks

4. Write these numbers as words.

    a) 3 _____   **b)** 10 _____   **c)** 20 _____

3 marks

PS **5.** Write the numbers.

    a) Seven _____   **b)** Twelve _____   **c)** Nine _____

3 marks

6. These toy trucks are different lengths.

A	B	C	D
5cm	8cm	10cm	3cm

    a) Which truck is the **shortest**? _____

    b) Which truck is the **longest**? _____

    c) Is truck D **longer** or **shorter** than A? _____

    d) Put the trucks in order from shortest to longest.

      shortest   longest

    e) If all of the trucks were put in a line, what would this measure in cm? _____ cm

5 marks

7. What would 1 more and 1 less of these numbers be?

    a)

	← 1 less	15	1 more →	

    b)

	← 1 less	20	1 more →	

    c)

	← 1 less	45	1 more →	

3 marks

8. Write 5 less and 5 more than the numbers shown.

a)

	← 5 less	50	5 more →	

b)

	← 5 less	35	5 more →	

c)

	← 5 less	20	5 more →	

3 marks

9. Draw the hands on the clocks to show the following times.

a)

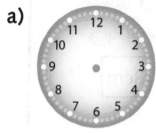

3 o'clock

b)

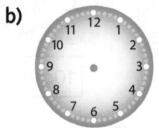

Half past 4

c)

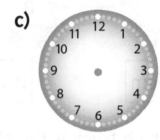

6 o'clock

3 marks

10. Write an addition number sentence and a subtraction number sentence for each set of numbers.

> **Example:** 2 3 5   2 + 3 = 5   3 + 2 = 5   5 − 3 = 2

a)

 6   2   8   ___ + ___ = ___   ___ − ___ = ___

b)

 9   1   10   ___ + ___ = ___   ___ − ___ = ___

c)

 5   7   12   ___ + ___ = ___   ___ − ___ = ___

3 marks

11. Count the racing cars.

a) How many cars would be double this number? _____ cars

b) How many cars would half the amount be? _____ cars

2 marks

6

PS **12.** This is a piece of a round pie.

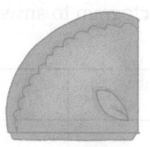

a) What fraction of the pie can you see? Tick your answer.

$\frac{1}{4}$ ☐ $\frac{1}{2}$ ☐ $\frac{3}{4}$ ☐

b) How much of the pie has been eaten? Tick your answer.

$\frac{1}{4}$ ☐ $\frac{1}{2}$ ☐ $\frac{3}{4}$ ☐

2 marks

13. Complete this number line counting in 5s.

	5		15	20				40		50

1 mark

14. Use the symbols to compare the numbers.

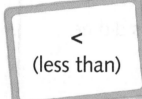   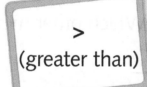   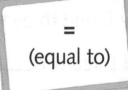

<   >   =
(less than)   (greater than)   (equal to)

Example: 23 < 45   76 > 53   81 = 81

a) 15 _____ 10   b) 19 _____ 20   c) 13 _____ 13

3 marks

7

PS **15.** Guy has made a pictogram of minibeasts that he found during a bug hunt! Use the pictogram to answer the questions.

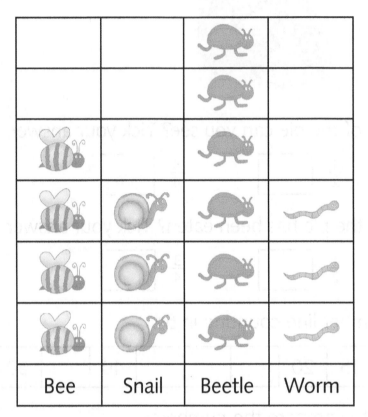

Bee	Snail	Beetle	Worm

a) How many beetles did Guy find? _____

b) Guy found three worms. Which other minibeasts did he find three of? _____

c) How many bees did Guy find? _____

d) How many minibeasts did Guy find in total? _____

4 marks

16. How much does the bowl weigh?

_____ g

1 mark

17. Answer these multiplication problems.

   a) 10 × 2 = _____   **b)** 5 × 5 = _____

   c) 3 × 10 = _____   **d)** 7 × 2 = _____

   e) 8 × 5 = _____   **f)** 6 × 10 = _____

6 marks

18. Count the money.

What is the total? _____p

 19.a) Count the busy bees. How many are there?

_____ bees

1 mark

b) If three bees flew away, how many would be left?

   _____ bees

c) Then, if seven bees joined the group, how many would there be?

   _____ bees

3 marks

PS **20.a)** Susan wants to share her pizza equally with a friend. What fraction do they each have? Tick one.

$\frac{1}{4}$ ☐

$\frac{1}{2}$ ☐

$\frac{3}{4}$ ☐

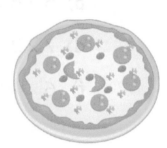

b) If Susan sliced her pizza into four equal parts, what fraction would each part be? Tick one.

$\frac{1}{4}$ ☐ $\frac{1}{2}$ ☐ $\frac{3}{4}$ ☐

2 marks

21. Write six addition number facts for 10.

Example: 10 + 0 = 10

_____ + _____ = 10   _____ + _____ = 10

_____ + _____ = 10   _____ + _____ = 10

_____ + _____ = 10   _____ + _____ = 10

6 marks

22. Write the name for each 2-D shape.

a) 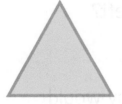   **b)**   **c)**

_____   _____   _____

3 marks

23. Tick the 3-D shape that is a cube.

 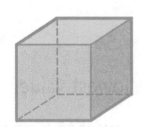 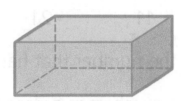

1 mark

24. Halve these numbers.

    a) Half of 10 is _____.   **b)** Half of 12 is _____.

    c) Half of 20 is _____.   **d)** Half of 8 is _____.

4 marks

25. Double these numbers.

    a) Double 3 is _____.   **b)** Double 5 is _____.

    c) Double 10 is _____.   **d)** Double 6 is _____.

4 marks

26. Follow the instructions and complete the grid.

    a) Draw a **circle** in the **centre** square.

    b) Put a **rectangle** in the square **above** the circle.

    c) Draw a **triangle** in the square **below** the circle.

3 marks

Marks........ /78

11

Numbers and Counting

1 Look at the numbers below.

45   11   21   56   33

a) Circle the number that has the lowest value.

b) Put the numbers in order from lowest to highest.

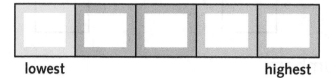

lowest   highest

2 marks

2 Look at the numbers below.

14   50   22   71   18

a) Circle the number that has the highest value.

b) Put the numbers in order from lowest to highest.

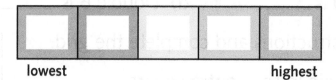

lowest   highest

2 marks

3 Write the missing numbers.

a)

	21			24	25		27		

b)

30	31				35		37		

2 marks

4 Write any number that comes between each pair of numbers.

a)

50		60

b)

45		61

2 marks

Marks.......... /8

12

Numbers and Counting

Challenge 2

1 Write these number words as numerals.

a) seventy-three _____ b) thirty-five _____

c) eighty-seven _____ d) fifty-six _____

e) seventeen _____ f) ninety-one _____

g) fifty-nine _____ h) twelve _____

i) twenty-two _____ j) one hundred and six _____

10 marks

Marks......... /10

Challenge 3

1 Write these numerals as number words in the train carriages.

a)  18

b) 97

c) 56

d) 15

e) 81

5 marks

Marks.......... /5

Total marks .... /23   How am I doing?

Counting Forwards and Backwards

PS Problem-solving questions

Challenge 1

Use this number line to help you to count.

11	12	13	14	15	16	17	18	19	20
21	22	23	24	25	26	27	28	29	30

1 Start at **27** and count **back**.

a) Count back 6 _____   b) Count back 13 _____

c) Count back 9 _____   d) Count back 16 _____

e) Count back 10 _____   f) Count back 11 _____

6 marks

2 Start at **12** and count **forwards**.

a) Count forwards 14 _____   b) Count forwards 11 _____

c) Count forwards 7 _____   d) Count forwards 15 _____

e) Count forwards 17 _____   f) Count forwards 3 _____

6 marks

Marks.........../12

Challenge 2

 1 Now solve these problems.

a) 26 fish are swimming in a pond. 5 fish swim away and hide in some weeds.

How many fish are still swimming in the pond? _____

b) 6 more fish join the ones that are left. How many fish are there now? _____ fish

Counting Forwards and Backwards

c) Cole has 51 stickers on his sticker sheet. He gives his friends 9 of the stickers. How many stickers does Cole have left? _____ stickers

d) Molly collects badges. She has 35 badges. Molly gets 7 more badges. How many does she have now?

_____ badges

e) There are 50 straws in a box. 9 straws are used.

How many straws are left? _____ straws

5 marks

2 Fill in the missing numbers.

1		3		5		7		9	
11		13		15		17		19	

1 mark

Marks.......... /6

Challenge 3

1
 62 34 45 89 23 76 13

Put these numbers in order from lowest to highest.

lowest | | | | | | | highest

1 mark

2 Start at **50** and count **back**.

a) Count back 12 _____   **b)** Count back 20 _____

c) Count back 19 _____   **d)** Count back 15 _____

4 marks

Marks.......... /5

Total marks .... /23   How am I doing?

Counting in Steps of 2, 3, 5 and 10

PS Problem-solving questions

Challenge 1

1 Continue the numbers counting in steps of 2. Be careful, some go backwards!

a)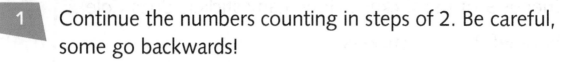

b)

c)

3 marks

2 Look at the following sequences and write down the number of steps they count in.

a) 45, 47, 49, 51

The numbers are counting in steps of _____.

b) 36, 46, 56, 66

They are counting in steps of _____.

2 marks

Marks............/5

Challenge 2

1 Fill in the blank flowers. Make sure you count in steps of 5!

a)

b)

c)

3 marks

16

Counting in Steps of 2, 3, 5 and 10

2 These numbers count in steps of 5. What are the next three steps?

a) 61, 56, 51, 46   The next three steps are ＿＿ ＿＿ ＿＿.

b) 34, 39, 44, 49   The next three steps are ＿＿ ＿＿ ＿＿.

c) 99, 94, 89, 84   The next three steps are ＿＿ ＿＿ ＿＿.

3 marks

Marks.......... /6

Challenge 3

 1 Fill in the missing number columns to complete them.

a)
58
43
23

b)
75
90
100

c)
26
21
16

d)
37
32
27

4 marks

 2 Write down if these numbers count in steps of **3** or **5**.

a) 23, 20, 17, 14, 11   These count in steps of ＿＿.

b) 34, 39, 44, 49, 54   These count in steps of ＿＿.

c) 62, 65, 68, 71, 74   These count in steps of ＿＿.

3 marks

Marks.......... /7

Total marks .... /18   How am I doing?

Counting in Steps of 2, 3, 5 and 10

 PS Problem-solving questions

Challenge 1

PS **1** Answer these problems.

a) Each pencil case holds 10 pencils.

Count the total number of pencils in lots of 10.

There are _____ pencils in total.

b) How many lots of 10 are there in 40? _____ lots

c) Complete the grid. Make sure you are counting in steps of 10!

	40			70			100

d) Start at 60 and count forwards 3 steps of 10. Which number are you at? _____

e) If 1 person has 10 toes, how many toes do 3 people have? _____ toes

f) How many toes would 5 people have? _____ toes

g) How many people would 40 toes belong to? _____ people

7 marks

Marks.......... /7

Challenge 2

1 Put the numbers in order to make a sequence counting **forwards** in steps of 10.

45, 75, 35, 15, 5, 65, 55, 25

1 mark

Counting in Steps of 2, 3, 5 and 10

2 Put the numbers in order to make a sequence counting **backwards** in steps of 10.

20, 50, 70, 60, 10, 30, 80, 40

1 mark

PS **3** Each ladybird has three spots. The spots are counted in steps of 3.

a) Continue to count in steps of 3.

3, 6, 9, 12 _____, _____, _____, _____

b) Now try this.

24, 21, 18, 15 _____, _____, _____, _____

c) How many ladybirds would 6 spots belong to?

_____ ladybirds

3 marks

Marks.......... /5

Challenge 3

PS **1** Complete this number grid. Check the steps each line counts in. Check whether they count forward or backward!

6		12		18
21		51		
	16		20	
67		61		55

4 marks

Marks.......... /4

Total marks .... /16   How am I doing?

Counting More and Less

Challenge 1

PS **1** Find 1 more and 1 less of these numbers.

	65			78			19	

	99			120			47	

6 marks

2 What is 5 more than each number?

a) 26 _____   **b)** 34 _____

c) 65 _____   **d)** 95 _____

e) 115 _____   **f)** 88 _____

6 marks

Marks........../12

Challenge 2

PS **1** Which of these numbers is bigger?

a) 23 or 17? _____   **b)** 67 or 81? _____

c) 77 or 76? _____   **d)** 11 or 13? _____

4 marks

2 Which of these numbers is smaller?

a) 76 or 56? _____   **b)** 99 or 88? _____

c) 101 or 107? _____   **d)** 34 or 17? _____

4 marks

3 Write the number that is 3 more than 39 _____

1 mark

Marks........../9

Counting More and Less

Challenge 3

PS **1** Write the number that is 10 less.

a) 44 _____   b) 76 _____   c) 110 _____

d) 152 _____   e) 194 _____   f) 243 _____

6 marks

PS **2** Answer these number problems.

a) Ryan has 16 games for his computer. Hassan has 7 more than Ryan. How many games does Hassan have?

_____ games

b) Joe has 2 tomato plants. He collects 12 tomatoes from the first plant. The second tomato plant has 8 fewer than the first. How many tomatoes does the second plant have?

_____ tomatoes

2 marks

PS **3** Lizzie takes 20 buns to school for the bun sale. Amber takes 5 fewer than Lizzie, and Tyler takes 7 more than Lizzie.

a) How many buns did Amber bring? _____ buns

b) Tyler brought _____ buns.

c) How many buns did they bring altogether? _____ buns

3 marks

Marks.......... /11

Total marks .... /32   How am I doing?

Place Value

PS Problem-solving questions

Challenge 1

1 Write the number that each abacus shows.

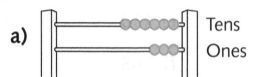

a) [abacus] Tens / Ones

b) [abacus] Tens / Ones

_____

c) [abacus] Tens / Ones

d) [abacus] Tens / Ones

_____   _____

4 marks

2 Draw beads on each abacus to show these numbers.

a) 26 [abacus] Tens / Ones

b) 35 [abacus] Tens / Ones

c) 76 [abacus] Tens / Ones

d) 89 [abacus] Tens / Ones

4 marks

Marks.......... /8

Challenge 2

1 Partition these two-digit numbers into tens and ones.

Example: $46 = 40 + 6$

a) $82 =$ _____ + _____

b) $15 =$ _____ + _____

c) $98 =$ _____ + _____

d) $36 =$ _____ + _____

4 marks

Place Value

2 Write the two-digit number made with these tens and ones.

Example: 60 + 2 = 62

a) 60 + 5 = _____ b) 50 + 1 = _____ c) 70 + 2 = _____

d) 40 + 7 = _____ e) 10 + 9 = _____ f) 30 + 6 = _____

 6 marks

Marks......... /10

Challenge 3

PS 1 My number is between 21 and 34. Write **one** of the numbers that my number could be. _____

1 mark

2 A jar has between 80 and 100 marbles. What do you **estimate** could be the true number? _____

1 mark

3 How many tens and ones do these two-digit numbers have?

a) 34: _____ tens, _____ ones

b) 84: _____ tens, _____ ones

c) 66: _____ tens, _____ ones

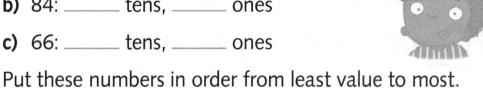

3 marks

4 Put these numbers in order from least value to most.

Example: 7, 13, 10, 15 would be 7, 10, 13, 15.

a) 12 1 7 21 ⬡ ⬡ ⬡ ⬡

b) 45 27 98 36 ⬡ ⬡ ⬡ ⬡

2 marks

Marks......... /7

Total marks .... /25   How am I doing?

Less Than, Greater Than and Equal To

 PS Problem-solving questions

Challenge 1

 PS **1** Put the numbers in pairs.

(23) (14) (23) (10) (73) (66) (73) (10) (12) (54)

a) _____ is less than _____

b) _____ is greater than _____

c) _____ is equal to _____

d) _____ is greater than _____

4 marks

Marks.......... /4

Challenge 2

1 Use symbols to compare the numbers.
< (less than)   > (greater than)   = (equal to)

Example: 23 < 45, 76 > 53, 81 = 81

a) 67 [ ] 98   **b)** 93 [ ] 82   **c)** 17 [ ] 19

3 marks

2 Use symbols to compare the numbers written in words.
< (less than)   > (greater than)   = (equal to)

Example: thirty-five < fifty-one

a) eighty-seven [ ] ninety-two

b) thirty-one [ ] twenty-five

c) twenty-seven [ ] twenty-seven

3 marks

Marks.......... /6

Less Than, Greater Than and Equal to

Challenge 3

PS **1** Complete the grid with symbols and numbers.

15		15
	>	
54		84
	=	
17		15

PS **2** Answer the following questions using words.

a) The number of oranges is _____ the number of apples.

b) There are _____ pears than strawberries.

c) The number of pears is _____ the number of oranges.

d) Which three have equal amounts? _____

_____

Marks.......... /9

How am I doing?

Doubling and Halving using Addition and Subtraction

PS Problem-solving questions

Challenge 1

PS **1** Double and halve the numbers on the three sets of balloons.

Example: 5, 10, 5   Double 5 is 10, half of 10 is 5

a) 10 20 10   **b)** 15 30 15   **c)** 100 100 200

a) _____

b) _____

c) _____

3 marks

2 Now try doubling and halving these numbers.

Example: Double 3 = 3 + 3 = 6, half of 6 = 3

a) 12 _____

b) 20 _____

c) 200 _____

3 marks

Marks.......... /6

Challenge 2

PS **1** Calculate and write these sums in a different order.

Example: 2 + 3 = 5, 3 + 2 = 5

a) 15 + 5 = _____   **b)** 23 + 10 = _____

_____ + _____ = _____   _____ + _____ = _____

c) 34 + 12 = _____   **d)** 54 + 14 = _____

_____ + _____ = _____   _____ + _____ = _____

4 marks

Doubling and Halving using Addition and Subtraction

2 Complete the fact families for these sets of numbers.

Example: You can use 2, 3 and 5 as follows:
2 + 3 = 5, 3 + 2 = 5, 5 – 3 = 2, 5 – 2 = 3

a) 15   20   35   **b)** 45   15   60

c) 30   70   100   **d)** 98   1   99

a) _____ _____ _____ _____

b) _____ _____ _____ _____

c) _____ _____ _____ _____

d) _____ _____ _____ _____

4 marks

Marks.......... /8

Challenge 3

1 Solve the following addition number sentences.
Remember to add the tens and then the units (ones).

a) 56 + 33 + 2 = _____   **b)** 61 + 16 + 5 = _____

c) 34 + 33 + 6 = _____   **d)** 75 + 13 + 10 = _____

4 marks

PS **2** Solve the following subtraction number sentences.
Remember to count back the number of tens in the smallest number and then the units.

a) 65 – 31 = _____   **b)** 89 – 36 = _____

c) 44 – 11 = _____   **d)** 75 – 14 = _____

4 marks

Marks.......... /8

Total marks .... /22   How am I doing?

Solving Number Problems

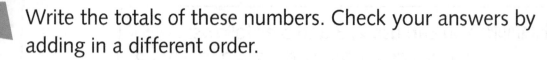

PS ⟩ Problem-solving questions

Challenge 1

1 Write the totals of these numbers. Check your answers by adding in a different order.

a) (5) (7) (1)

_____ + _____ + _____ = _____

_____ + _____ + _____ = _____

b) (3) (10) (2)

_____ + _____ + _____ = _____

_____ + _____ + _____ = _____

c) (6) (11) (5)

_____ + _____ + _____ = _____

_____ + _____ + _____ = _____

d) (14) (6) (3)

_____ + _____ + _____ = _____

_____ + _____ + _____ = _____

4 marks

Marks............/4

Challenge 2

1 Answer these addition and subtraction problems. You could draw pictures to help.

a) Martin has 15 monsters. He gives 6 monsters to his friends. How many monsters does Martin have left?

_____ – _____ = _____

b) Sita has collected 12 shells. Her sister gives her 11 more shells. How many shells does Sita have altogether?

_____ + _____ = _____

Solving Number Problems

c) Helen has 20 beads. She gives 10 to her friend Liz.
How many beads does Helen have now?

_____ – _____ = _____

d) David gets 26 pens in a pack. He gives Zac 9 of them.
How many pens does David have now?

_____ – _____ = _____

4 marks

Marks........... /4

Challenge 3

PS **1** Rita has 3 boxes of apples. Each box holds
20 apples. How many apples does Rita

have in total? _____ apples

Show this as an addition sum.

_____ + _____ + _____ = _____

2 marks

2 Fatima collects 22 conkers. Andrew finds 15 conkers.

How many more conkers does Fatima have? _____ conkers

Show how you worked this out.

_____ – _____ = _____

2 marks

3 There are 29 children in class four and 30 in class five.

How many children are there in total? _____ children

2 marks

Marks........... /6

Total marks .... /14

How am I doing?

Using Two–Digit Numbers

Challenge 1

PS | **1** | There are 20 biscuits on a plate.
Write nine addition number facts
for the total number of biscuits.

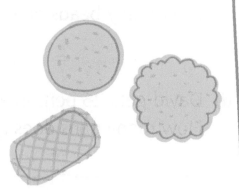

Example: 19 + 1 = 20

9 marks

Marks............/9

Challenge 2

1 | Write ten different two-digit numbers using the numbers
on the cards.

_____   _____   _____   _____   _____

_____   _____   _____   _____   _____

10 marks

2 | Add 10 to each of these two-digit numbers.

a) 23 _____   **b)** 54 _____   **c)** 36 _____

d) 16 _____   **e)** 78 _____   **f)** 47 _____

6 marks

Marks........../16

Using Two-Digit Numbers

Challenge 3

PS **1** There are 20 flowers left in a field after some have been picked.

Write nine subtraction number facts to show how many flowers could have been in the field.

Example: 26 – 6 = 20

9 marks

PS **2** The lily pads are numbered with two-digit numbers. Write the numbers on the blank lily pads so that they are in the correct sequence. Be careful to work out what steps to count in!

26   36   51

4 marks

Marks......... /13

Total marks .... /38   How am I doing?

31

Solving Missing Number Problems

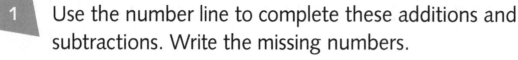

 Problem-solving questions

Challenge 1

1 Use the number line to complete these additions and subtractions. Write the missing numbers.

```
0        5        10       15       20       25       30       35       40
├┼┼┼┼┼┼┼┼┼┼┼┼┼┼┼┼┼┼┼┼┼┼┼┼┼┼┼┼┼┼┼┼┼┼┼┼┼┼┼┼┤
```

a) $12 + \underline{\hphantom{xxx}} = 20$ **b)** $24 + \underline{\hphantom{xxx}} = 32$

c) $\underline{\hphantom{xxx}} + 10 = 32$ **d)** $19 + \underline{\hphantom{xxx}} = 29$

e) $\underline{\hphantom{xxx}} + 16 = 20$ **f)** $25 - \underline{\hphantom{xxx}} = 18$

g) $20 - 8 = \underline{\hphantom{xxx}}$ **h)** $14 - \underline{\hphantom{xxx}} = 3$

8 marks

Marks.......... /8

Challenge 2

 **1** Casey is selling bows at the school fair, but she has lost her price tags! Help her to remember the price of her bows and fill in the missing amounts. Casey knows that the purple bows cost 10p.

a) The pink bows cost 3p more

than the purple bows. Pink bows cost _____p.

b) Blue bows cost 10p more than the purple bows.

Blue bows cost _____p.

c) Casey's green bows cost 2p more than the blue bows.

Green bows cost _____p.

Solving Missing Number Problems

d) A yellow bow costs the same as a pink bow and a blue bow added together. A yellow bow costs _____ p.

e) An orange bow costs 5p less than a purple bow.

An orange bow costs _____ p.

5 marks

Marks.......... /5

Challenge 3

1 Fill in the missing number operation symbols (– or +).

a) 30 _____ 70 = 100

b) 25 _____ 15 = 10

c) 35 _____ 10 = 45

d) 47 = 67 _____ 20

e) 50 = 35 _____ 15

5 marks

2 Complete this number grid.

25		27		
	36			39
		47		
55			58	
	66			69

5 marks

Marks......... /10

Total marks .... /23

How am I doing?

PS Problem-solving questions

1. Complete this addition grid. Make sure that each column and row has a total of 12. You can use the same number more than once.

1		
	3	
		8

3 marks

2. Make ten different two-digit numbers using these single digit numbers.

Example: Digits 1 and 3 make 13

10 marks

PS 3. Answer these number problems.

a) Savi has 6 caterpillars. He collects 7 more. How many does Savi have now? _____ caterpillars

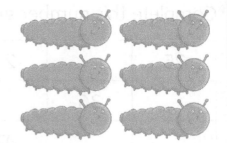

b) The 7 caterpillars escape and Savi is back to 6. His friend Jemma takes three of them.

How many caterpillars remain? _____ caterpillars

c) Each of Jemma's 3 caterpillars finds 5 friends. How many caterpillars does she have in total?

_____ caterpillars

3 marks

4. Write each addition in a different order.

a) 27 + 10 _____ **b)** 32 + 87 _____

c) 164 + 34 _____ **d)** 76 + 120 _____

4 marks

5. Circle the number that has the highest value.

19   80   33   21   79

1 mark

6. Write these as numerals.

a) twenty-seven _____ **b)** ninety-four _____

c) sixteen _____ **d)** thirty-nine _____

e) fourteen _____ **f)** seventy-six _____

6 marks

PS **7.** Tommy has some big and small blocks. The big blocks are worth 10 and the small blocks are worth 1. Calculate the numbers that Tommy has made using his blocks.

a)  _____

b)  _____

_____

c) _____

3 marks

8. Write 5 less than each number.

a) 17 _____   b) 10 _____   c) 127 _____

d) 76 _____   e) 38 _____   f) 101 _____

6 marks

9. Write these numbers as words.

a) 50 _____

b) 17 _____

c) 89 _____

d) 31 _____

e) 73 _____

5 marks

10. Double these numbers.

a) 25 _____   b) 34 _____   c) 50 _____

d) 13 _____   e) 27 _____   f) 42 _____

6 marks

11. What would half of these numbers be?

a) 12 _____   b) 24 _____   c) 100 _____

d) 32 _____   e) 16 _____   f) 78 _____

6 marks

12.a) Use any two numbers from the different cards. Write the lowest possible two-digit number. _____

b) Write the highest possible number. _____

c) Write the highest three-digit number you can make. _____

d) Write the lowest three-digit number you can make. _____

4 marks

PS **13.** Write addition and subtraction fact families for these sets of numbers.

> **Example:** 13 + 7 = 20, 20 – 7 = 13

a) 25 30 55   **b)** 140 60 200

c) 31 9 40   **d)** 38 52 90

a) _____ _____ _____ _____

b) _____ _____ _____ _____

c) _____ _____ _____ _____

d) _____ _____ _____ _____

4 marks

14. Write 5 less than each number.

a) 77 ___ **b)** 100 ___ **c)** 127 ___ **d)** 76 ___ **e)** 38 ___

5 marks

15. Henry has 17 marbles. He wins 6 more. How many marbles does Henry have now? _____ marbles

1 mark

16.a) How many birds are there in total? _____ birds

b) How many lots of 2 are there? _____ lots

2 marks

Marks.........../69

What is Multiplication?

Challenge 1

PS **1** Count the oranges in lots of 5.

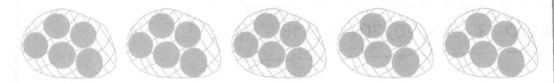

a) There are _____ oranges in total.

b) Write this as an addition.

_____ + _____ + _____ + _____ + _____ = _____

c) Now write it as a multiplication.

_____ × _____ = _____

3 marks

2 Now write the multiplication for these additions.

a) 2 + 2 + 2 = 6 _____ × _____ = _____

b) 5 + 5 = 10 _____ × _____ = _____

c) 3 + 3 + 3 + 3 = 12 _____ × _____ = _____

3 marks

3 How much money is there altogether?

_____ + _____ + _____ + _____ + _____ + _____ +

_____ + _____ + _____ + _____ = £ _____

1 mark

Marks.........../7

38

What is Multiplication?

Challenge 2

PS | **1** There are two possible multiplications shown by this array. What are they?

_____ × _____ = _____ and _____ × _____ = _____

2 Draw an array for each of these multiplications. Use a separate piece of paper.

a) 3 × 4   **b)** 4 × 2   **c)** 2 × 5   **d)** 3 × 10

Marks.......... /6

Challenge 3

1 Complete the multiplication table. The first one has been done for you.

Repeated Addition	Multiplication 1	Multiplication 2
3 + 3 + 3 + 3 = 12	3 × 4 = 12	4 × 3 = 12
	5 × 3 = 15	
10 + 10 + 10 = 30		
		4 × 5 = 20
	5 × 7 = 35	
2 + 2 + 2 + 2 + 2 = 10		

Marks.......... /10

Total marks .... /23   How am I doing?

What is Division?

Challenge 1

 1 Answer these division problems.

a) You and your friend have ten badges. Divide them so that you both have the same amount. How many badges would you each get?

Each person would get _____ badges.

b) If you shared the badges between five, how many would you each get now?

Each person would get _____ badges.

c) How many badges would you need for five people to each have five badges?

You would need _____ badges.

3 marks

Marks........../3

Challenge 2

 1 Share these buns equally between two people. Draw the buns on the plates.

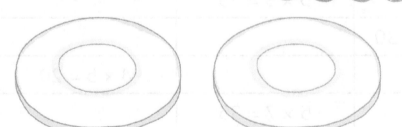

a) Each person will get _____ buns.

b) Write it as a division sentence. _____ ÷ _____ = _____

2 marks

What is Division?

2 Share these 16 sweets equally so everyone gets the same amount.

a) Each person will get

_____ sweets.

b) Write this as a division sentence.

_____ ÷ _____ = _____

2 marks

3 Share 15 watermelon slices among three people. How many would each person get?

a) Each person would get _____ watermelon slices.

b) Write this as a division sentence. _____ ÷ _____ = _____

2 marks

Marks.......... /6

Challenge 3

1 Complete this division table. The first one has been done for you.

Numbers of Items	Number of People	Division Sentence
20	4	20 ÷ 4 = 5
10	2	
		12 ÷ 3 = 4
20	4	
18	3	
		40 ÷ 2 = 20

7 marks

Marks.......... /7

Total marks .... /16   How am I doing?

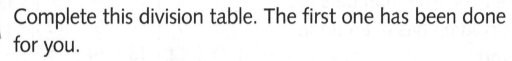

2, 5 and 10 Multiplication Tables

Challenge 1

1 Complete this multiplication table. The first one has been done for you.

1 × 2 = 2	1 × 5 = 5	1 × 10 = 10
	3 × 5 = 15	3 × 10 = 30
4 × 2 = 8		
		5 × 10 = 50
	6 × 5 = 30	
7 × 2 = 14		7 × 10 = 70
	8 × 5 = 40	
9 × 2 = 18		9 × 10 = 90

17 marks

Marks......... /17

Challenge 2

1 Colour the even numbers **red**.
One column has been done
for you.

1	**2**	3	4	5
6	7	8	9	10
11	**12**	13	14	15
16	17	18	19	20
21	**22**	23	24	25

4 marks

2 Write **odd** or **even** for the number of eggs in each nest.

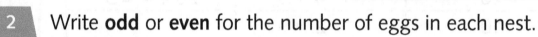

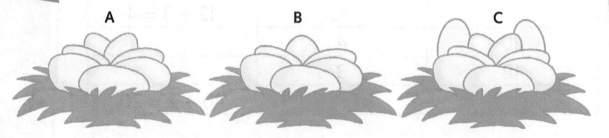

A   B   C

a) Nest A has an _____ number of eggs.

2, 5 and 10 Multiplication Tables

b) Nest B has an _____ number of eggs.

c) The third nest has an _____ number of eggs.

3 marks

3 Are these statements **true** or **false**?

a) When you multiply any number by 2 the answer is always an even number. _____

b) When you multiply a number by 5 the answer is always an odd number. _____

2 marks

Marks........../9

Challenge 3

1 Multiplication is **commutative**. It has the same **product** if the **numbers** are reversed.

Example: 3 × 2 = 6 and 2 × 3 = 6

Reverse these multiplications.

a) 3 × 5 = 15 _____

b) 2 × 10 = 20 _____

c) 10 × 5 = 50 _____

d) 9 × 2 = 18 _____

e) 3 × 10 = 30 _____

f) 5 × 10 = 50 _____

6 marks

Marks........../6

Total marks ..../32

How am I doing?

Division Problems

Challenge 1

PS **1** Look at the strawberries. Write down the division sentence that you can see.

_____ ÷ _____ = _____

1 mark

2 Use these numbers to write division sentences.

a) 15   5   3   _____ ÷ _____ = _____

b) 20   10   2   _____ ÷ _____ = _____

2 marks

3 Now write division sentences for these numbers.

a) 5   25   5   _____ ÷ _____ = _____

b) 40   8   5   _____ ÷ _____ = _____

2 marks

Marks.......... /5

Challenge 2

PS **1** Rashid has a collection of 30 stamps. He wants to share them with his friends.

a) If Rashid shared his stamps between two friends, how many would they get each? _____

Write this as a division sentence. _____ ÷ _____ = _____

Division Problems

b) If five friends shared the stamps, they would have

_____ each.

Write this as a division sentence. _____ ÷ _____ = _____

c) If Rashid shared his stamps between 10 friends, they

would each get _____ stamps.

Write this as a division sentence. _____ ÷ _____ = _____

6 marks

Marks.......... /6

Challenge 3

1 Jasmine has 30p in her money box.

a) If she has only 5p coins, how many coins does she

have? _____ coins

b) If the 30p was made up of 10p coins, how many

would there be? _____ coins

2 marks

2 Write the answers.

a) $50 \div 10 =$ _____

b) $30 \div 5 =$ _____

c) $16 \div 2 =$ _____

d) $60 \div 5 =$ _____

4 marks

3 Complete this division table.

20	÷	5	=	
14			=	7
	÷	10		10

5 marks

Marks.......... /11

Total marks .... /22   How am I doing?

Connecting Multiplication and Division

 **PS** Problem-solving questions

Challenge 1

1 Write the division sentence to check these multiplications.

a) $2 \times 6 = 12$ _____ ÷ _____ = _____

b) $5 \times 10 = 50$ _____ ÷ _____ = _____

c) $3 \times 5 = 15$ _____ ÷ _____ = _____

d) $2 \times 10 = 20$ _____ ÷ _____ = _____

e) $10 \times 10 = 100$ _____ ÷ _____ = _____

5 marks

2 Now write the multiplication sentences to check these divisions.

a) $15 \div 5 = 3$ _____ × _____ = _____

b) $35 \div 5 = 7$ _____ × _____ = _____

c) $20 \div 5 = 4$ _____ × _____ = _____

d) $20 \div 2 = 10$ _____ × _____ = _____

e) $90 \div 10 = 9$ _____ × _____ = _____

5 marks

Marks......... /10

Challenge 2

1 Use these numbers to make multiplication and division number sentences.

Example:

$6 \times 5 = 30 \quad 5 \times 6 = 30 \quad 30 \div 5 = 6 \quad 30 \div 6 = 5$

Connecting Multiplication and Division

a) (7) (5) (35)

_____ × _____ = _____

_____ × _____ = _____

_____ ÷ _____ = _____

_____ ÷ _____ = _____

b) (2) (10) (20)

_____ × _____ = _____

_____ × _____ = _____

_____ ÷ _____ = _____

_____ ÷ _____ = _____

8 marks

2 Write the sets of 3 numbers that were used for each multiplication and division number sentence.

a) $5 \times 9 = 45$ $9 \times 5 = 45$ $45 \div 9 = 5$ $45 \div 5 = 9$

_____, _____ and _____ were used.

b) $10 \times 6 = 60$ $6 \times 10 = 60$ $60 \div 6 = 10$ $60 \div 10 = 6$

_____, _____ and _____ were used.

2 marks

Marks......... /10

Challenge 3

 1 a) Divide this group of motorbikes by two.

What answer do you get? _____

Write this as a division sentence.

_____ ÷ _____ = _____

b) Multiply the bikes by two. The answer

is _____ bikes.

Write this as a multiplication sentence.

_____ × _____ = _____

4 marks

Marks.......... /4

Total marks .... /24   How am I doing?

Doubling and Halving using Multiplication and Division

 PS Problem-solving questions

Challenge 1

1 Multiply these numbers by 2 to double them.

> **Example:** 2 × 4 = 8   Double 4 is 8

a) 2 × 5 = _____   Double 5 is _____

b) 2 × 10 = _____   Double 10 is _____

c) 2 × 3 = _____   Double 3 is _____

d) 2 × 12 = _____   Double 12 is _____

4 marks

2 Now try these divisions to halve the numbers.

> **Example:** 10 ÷ 2 = 5   Half of 10 is 5

a) 12 ÷ 2 = _____   Half of 12 is _____

b) 16 ÷ 2 = _____   Half of 16 is _____

c) 20 ÷ 2 = _____   Half of 20 is _____

d) 14 ÷ 2 = _____   Half of 14 is _____

 4 marks

Marks.........../8

Challenge 2

 PS **1** Count these flowers.

a) How many flowers are there in

total? _____ flowers

b) How many flowers would half that

number be? _____ flowers

Write this as a division sentence. _____ ÷ _____ = _____

Doubling and Halving using Multiplication and Division

c) How many flowers would double that number be?

_____ flowers

Write this as a multiplication sentence.

_____ × _____ = _____

5 marks

PS 2 Now let us try the same with these children.

a) How many children are there in total? _____ children

b) How many children would half this group be? _____ children

Write this as a division sentence.

_____ ÷ _____ = _____

c) How many children would there be if you doubled the group? _____ children

Write this as a multiplication sentence.

_____ × _____ = _____

5 marks

Marks......... /10

Challenge 3

1 Complete these multiplication and division sentences.

6 × 2 = 12		12 ÷ 2 = 6	
	2 × 7 = 14		
		16 ÷ 2 = 8	
			20 ÷ 10 = 2
50 × 2 = 100			

14 marks

Marks......... /14

Total marks .... /32   How am I doing?

Solving Multiplication and Division Problems

PS ⟩ **Problem-solving questions**

Challenge 1

 1 Solve these real-life multiplication problems.

a) Ali had 2 hens. Each hen lays 1 egg every day. How many eggs did Ali's hens lay in 5 days?

_____ eggs

Write down how you worked this out using a

multiplication sentence. _____ × _____ = _____

2 marks

b) There are 22 sheep in a field. How many sheep would be half that number? _____ sheep

Show this as a division sentence. _____ ÷ _____ = _____

2 marks

Marks.........../4

Challenge 2

 1 This is class 3.

a) Count the children. There are _____ children.

b) They had to find partners. How many pairs of two can you count? _____ pairs

c) Write the division sentence to show this.

_____ ÷ _____ = _____

3 marks

 2 a) Jack had 15 pens. Lilly had double that number. How many pens did Lilly have? _____ pens

Solving Multiplication and Division Problems

b) Write down the calculation. _____

c) Did Jack have half the number of pens owned by Lilly? Circle the correct answer. YES / NO

3 marks

Marks.......... /6

Challenge 3

PS | **1** | Complete the table by answering the problems. The first one has been done for you.

Number Problem	× or ÷	Number Sentence
One dog has 4 legs. How many legs do 2 dogs have?	×	2 × 4 = 8
Two identical ladybirds have a total of 16 spots. How many spots does 1 ladybird have?		
There are 9 boys in class 5. Class 6 has double the number of boys. How many boys are in class 6?		
Ellie needs 40 straws for her party. They come in packs of 10. How many packs does Ellie need?		
Joe saves 50p each week. How many weeks will he have to save to have £1?		

8 marks

2 Make up your own number problem in the empty space in the table, and complete it.

3 marks

Marks.......... /11

Total marks .... /21   How am I doing?

51

Solving Multiplication and Division Problems

PS Problem-solving questions

Challenge 1

PS 1 Answer these multiplication and division problems.

a) A shop has 20 bouncy balls. They fit into boxes of 5. How many boxes will the shopkeeper need?

_____ boxes

Show the calculation. _____

b) Flowers are sold in bunches of 5. There are 10 bunches of flowers. How many flowers are there altogether?

_____ flowers

Show how you worked out the answer. _____

c) 50 children are eating lunch in the dining hall. Each table holds 10 children. How many tables are there?

_____ tables

How many children would be able to eat if there were only 3 tables? _____ children

Show how you worked this out. _____

7 marks

Marks.........../7

Challenge 2

PS 1 One ice cream costs 20p. Hildi has 50p. How many ice creams could she buy? _____ ice creams

Show the sum. _____

How much money would Hildi have left? _____p

3 marks

PS 2 John has 15 fish in his tank. He buys more to double this number. How many fish does John have now? _____ fish

Solving Multiplication and Division Problems

Write the multiplication sentence to show this.

_____ × _____ = _____

Now show the number sentence to halve the total.

_____

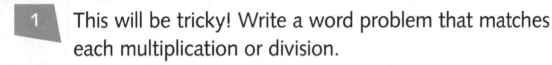 PS 3 Mia's rabbit, Snowy, eats 5 carrot snacks each day. How

many snacks does Snowy eat in 5 days? _____ snacks

How many snacks would Snowy eat in

10 days? _____ snacks

Mia buys the snacks in packs of 20. How many

days does one pack last? _____ days

Show how you worked this out. _____

Marks......... /10

Challenge 3

1 This will be tricky! Write a word problem that matches each multiplication or division.

5 × 4 = 20 _____

_____

12 ÷ 2 = 6 _____

_____

Marks.......... /2

Total marks .... /19   How am I doing?

What is a Fraction?

PS Problem-solving questions

Challenge 1

PS **1** Shade in one half of each square.

A   B   C   D

4 marks

2 How many separate halves are there in four squares?

_____ halves

1 mark

Marks.......... /5

Challenge 2

PS **1** Find a different way to halve each circle. Be very careful to make both sides equal or they are not proper fractions!

10 marks

Marks......... /10

What is a Fraction?

PS

1 Draw a line on each object to make two halves that look the same in a mirror. You could use a mirror to help.

A   B   C   D   E

5 marks

2 Draw the missing half of each object. Try very carefully to make the drawing as close to the half shown as possible. Your mirror could help again!

4 marks

Marks.......... /9

Total marks .... /24   How am I doing?

What is a Fraction?

Challenge 1

 1 Colour $\frac{1}{2}$ of each shape.

3 marks

2 Colour $\frac{1}{4}$ of each shape. Choose a different way each time.

3 marks

3 Colour $\frac{3}{4}$ of each shape. Choose a different way each time.

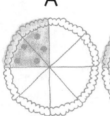   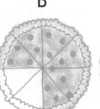

3 marks

Marks.........../9

Challenge 2

1 **a)** Draw lines to match the fraction with the correct pizza.

A   B   C   D   E

| 0 | $\frac{1}{4}$ | $\frac{1}{2}$ | $\frac{3}{4}$ | 1 whole |

What is a Fraction?

b) What fraction of each pizza has been eaten?

A = ☐   B = ☐   C = ☐

D = ☐   E = ☐

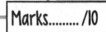
10 marks

Marks......... /10

Challenge 3

1 How many quarters do these fractions equal?

a) $\frac{1}{4}$ _____   **b)** $\frac{3}{4}$ _____

c) $\frac{1}{2}$ _____   **d)** 1 whole _____

4 marks

2 Complete this fractions table.

Whole	Halves	Quarters
1		
2		
4	8	
	6	
		24
	20	

11 marks

Marks......... /15

Total marks ..../34   How am I doing?

57

Fractions of Numbers

PS Problem-solving questions

Challenge 1

PS **1** Use $\frac{1}{4}$, $\frac{1}{2}$ or $\frac{3}{4}$ to answer these questions.

a) 6 is ☐ of 12

b) 2 is ☐ of 4

c) 6 is ☐ of 24

d) 50 is ☐ of 100

e) 30 is ☐ of 40

f) 20 is ☐ of 80

6 marks

2 Order these fractions from least to most.

$\frac{3}{4}$ $\frac{1}{4}$ 0 2 $\frac{1}{2}$

least ☐ ☐ ☐ ☐ ☐ most

1 mark

Marks.........../7

Challenge 2

PS **1** Write in the numbers that are the same as these fractions of larger numbers.

Example: $\frac{1}{2}$ of 10 is 5

a) $\frac{1}{2}$ of 20 is _____

b) $\frac{1}{4}$ of 20 is _____

c) $\frac{3}{4}$ of 20 is _____

d) $\frac{2}{4}$ of 20 is _____

e) $\frac{3}{4}$ of 100 is _____

f) $\frac{1}{4}$ of 100 is _____

6 marks

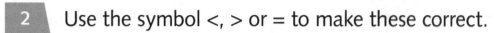

Fractions of Numbers

2 Use the symbol <, > or = to make these correct.

a) $\frac{1}{2}$ of 10 ☐ $\frac{1}{4}$ of 40

b) $\frac{3}{4}$ of 12 ☐ $\frac{1}{2}$ of 8

c) $\frac{2}{4}$ of 16 ☐ $\frac{1}{2}$ of 16

d) $\frac{1}{4}$ of 100 ☐ $\frac{1}{2}$ of 60

4 marks

Marks......... /10

Challenge 3

PS **1** Here are some whole numbers and fractions. Answer the questions about them.

Example: $1\frac{1}{2}$ = 3 halves or 6 quarters

a) How many halves are there in $2\frac{1}{2}$? _____ halves

b) How many quarters make $3\frac{1}{4}$? _____ quarters

c) $5\frac{1}{2}$ has how many quarters? _____ quarters

d) $10\frac{1}{4}$ has how many quarters? _____ quarters

e) $12\frac{3}{4}$ has how many halves? (Be careful with this one.)

_____ halves

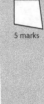

5 marks

2 Complete this fraction number line. It counts in quarters.

| $1\frac{1}{4}$ | $1\frac{1}{2}$ | $1\frac{3}{4}$ | 2 | | | | 3 | | |

5 marks

Marks......... /10

Total marks .... /27

How am I doing?

Finding Fractions of Larger Groups

PS Problem-solving questions

Challenge 1

PS **1** Find the fractions of these groups.

A  B  C

a) Which bag contains half the marbles of group B? _____

b) How many marbles would be half of group A?

_____ marbles

c) What would be $\frac{3}{4}$ of group C? _____ marbles

d) How many marbles would $\frac{1}{2}$ of group B be?

_____ marbles

e) What fraction of group C is 10 marbles? ☐

5 marks

2 Colour $\frac{1}{4}$ of each pizza and make each pattern different.
(Shaded sections do not have to be connected.)

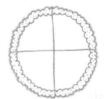 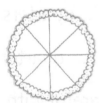 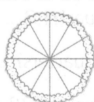 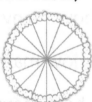

4 marks

Marks.........../9

Challenge 2

PS **1** Look at the things that you might expect to see at the seaside and answer the questions.

a) Draw a line around $\frac{1}{4}$ of the group.

Finding Fractions of Larger Groups

b) Draw a dotted line around half of the group.

c) Find $\frac{3}{4}$ of the group and draw a zig zag line around it.

3 marks

2 Here are four groups of 1p coins. Answer the questions about them.

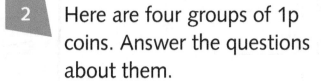

a) What would $\frac{1}{2}$ of the coins be worth? _____ p

b) What would $\frac{1}{4}$ of the coins be worth? _____ p

c) What would $\frac{3}{4}$ of the coins be worth? _____ p

3 marks

Marks.......... /6

Challenge 3

1 Complete the table.

Item	$\frac{1}{4}$	$\frac{1}{2}$	$\frac{3}{4}$
20 cm		10 cm	
60p			
40 teddies			
100 pencils			
80 flowers			

14 marks

Marks......... /14

Total marks .... /29

How am I doing?

1. Write the answers to these multiplications.

a) $3 \times 5 =$ _____

b) $8 \times 10 =$ _____

c) $7 \times 3 =$ _____

d) $9 \times 2 =$ _____

e) $10 \times 5 =$ _____

f) $4 \times 3 =$ _____

6 marks

2. Partition the two-digit numbers into tens and ones.

a) $87 =$ _____ tens, _____ ones

b) $59 =$ _____ tens, _____ ones

c) $41 =$ _____ tens, _____ ones

d) $60 =$ _____ tens, _____ ones

e) $99 =$ _____ tens, _____ ones

5 marks

PS **3.**

a) How many players would half of this team be? _____ players

b) How many players would there be in $\frac{1}{4}$ of the team?

_____ players

c) $\frac{3}{4}$ of the team would be _____ players.

d) How many players make up the whole team? _____ players

4 marks

4. Find half of these numbers by dividing.

   a) Half of thirty is _____   **b)** Half of twenty-four is _____

   c) Half of 18 is _____   **d)** Half of 50 is _____

   e) Half of 100 is _____   **f)** Half of sixty is _____

6 marks

5. Double these numbers by multiplying.

   a) Double 10 is _____   **b)** Double 20 is _____

   c) Double 50 is _____   **d)** Double 45 is _____

   e) Double 17 is _____   **f)** Double 12 is _____

6 marks

6. Colour these arrays to show the following calculations.

   a) $3 \times 2 = 6$ **b)** $6 \times 2 = 12$

2 marks

PS 7. Hopscotch the rabbit eats five carrot snacks every day. How many carrot snacks would he eat in:

   a) 20 days? _____ snacks

   b) 100 days? _____ snacks

2 marks

8. a) Count back using lots of 3 to complete these flowerpots.

39   30   24

1 mark

b) Count in 5s to complete this number tower.

45
35
15
10

4 marks

9. Show 5 less and 5 more of these numbers.

5 less		5 more
	43	
	51	
	29	
	33	
	87	

5 marks

10. Write the missing symbol <, > or =.

a) 78 ☐ 67   b) 77 ☐ 77

c) 89 ☐ 98   d) 13 ☐ 15

e) 23 ☐ 23   f) 65 ☐ 56

6 marks

11. Write the totals of these numbers. Check your answers by adding in a different order.

a)

3   9   2

_____ + _____ + _____ = _____

_____ + _____ + _____ = _____

b)

_____ + _____ + _____ = _____

_____ + _____ + _____ = _____

c)

_____ + _____ + _____ = _____

_____ + _____ + _____ = _____

d)

_____ + _____ + _____ = _____

_____ + _____ + _____ = _____

e)

_____ + _____ + _____ = _____

_____ + _____ + _____ = _____

5 marks

PS **12.** Find the fractions of these groups.

A   B   C

a) Which group contains half as many apples as group A? _____

b) How many apples are in $\frac{3}{4}$ of group B? _____ apples

c) How many apples are in $\frac{1}{2}$ of group C? _____ apples

d) How many apples are in $\frac{1}{2}$ of group A? _____ apples

e) What fraction of group C is 3 apples? ☐

5 marks

Marks......../57

65

Measuring Length and Height

Challenge 1

PS **1** These slippery snakes are different lengths.

A   B   C   D

 = 10 cm   = 15 cm   = 6 cm   = 20 cm

a) Which snake is the **longest**? _____

b) Which snake is the **shortest**? _____

c) Is snake B **shorter** or **longer** than snake C? _____

d) Put the snakes in order from shortest to longest.

_____ _____ _____ _____

e) If the snakes were joined together, what would they measure, in cm? _____ cm

5 marks

PS **2** **a)** Look at Gertie the giraffe. Write her height, in metres. _____ m

b) Her brother, Geoff, is exactly 1 m taller than Gertie. How tall is Geoff? _____ m

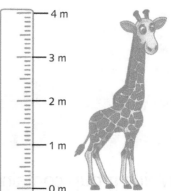

c) Gertie has a baby, she's called Georgia. Georgia is exactly half the height of Gertie. What is the height of Georgia? _____ m

d) Gertie's granddad, Grindle, was the tallest giraffe ever measured. He was 6 metres from hoof to horn!

How much taller than Gertie was Grindle? _____ m

4 marks

Marks.......... /9

Measuring Length and Height

Challenge 2

1 Use a cm ruler to measure these lines.

a) _____ = _____ cm

b) _____ = _____ cm

c) _____ = _____ cm

d) _____ = _____ cm

e) _____ = _____ cm

5 marks

Marks.......... /5

Challenge 3

1 Estimate, or best guess, the length of these shoelaces using whole numbers. The clues will help you.

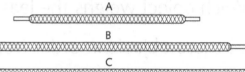

a) Which two shoelaces are about the same length?

_____ and _____

b) Shoelace A measures more than 1 cm, but less than 4 cm. It could be _____ cm.

c) Lace B is longer than 5 cm but shorter than 8 cm. It could measure _____ cm.

d) Lace C is the longest. It measures between 10 cm and 15 cm. It might be _____ cm.

e) Lace D is between 3 cm and 7 cm. What do you estimate its length to be? _____ cm

5 marks

Marks.......... /5

Total marks .... /19

How am I doing?

67

Measuring Weight and Volume

Challenge 1

 1

a) Draw lines to join each object to its estimated weight.

A Apple   Less than 20 g

B Washing machine   About 100 g

C Pencil   More than 20 kg

D Bag of potatoes   About 3 kg

b) Which object weighs the least? _____

c) Put these objects in order from lightest to heaviest.

_____ _____ _____ _____

 3 marks

Marks........../3

Challenge 2

1 These jugs contain different amounts of water. The scale measures in steps of 100 ml.

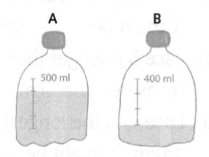

A   B
500 ml   400 ml

a) What is the volume of water

in jug A? _____ ml

b) How much water is in jug B? _____ ml

c) If you poured both amounts into one jug, what would

be the total volume? _____ ml

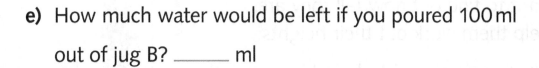

Measuring Weight and Volume

d) If you poured 100 ml out of jug A, how much water would be left? _____ ml

e) How much water would be left if you poured 100 ml out of jug B? _____ ml

5 marks

Marks.......... /5

Challenge 3

PS **1** Salima and her friends have been collecting cherries.

Freddie = 300 g   Salima = 250 g   Scott = 900 g   Rhona = 500 g

a) Who collected the **most**? _____

b) Who collected the **least**? _____

c) Does Salima's container have **more** or **less** than Rhona's container? _____

d) Who has collected twice the amount of Salima? _____

e) Put the friends in order according to how much they collected, from **most** to **least**.

_____ _____ _____ _____
   most   least

5 marks

Marks.......... /5

Total marks .... /13   How am I doing?

Comparing Measurements

Challenge 1

PS **1** Billy and his two brothers, Jack and Ben, want to find out how tall they are. Help them work out their heights.

Billy has measured Jack and he is exactly 64 cm tall.

a) Billy is twice the height of Jack,

so Billy must measure _____ cm.

b) Little Ben is 20 cm shorter than Jack. Ben measures

_____ cm.

c) Put the family in order from smallest to tallest.

_____   _____   _____

     smallest   tallest

3 marks

Marks.........../3

Challenge 2

PS **1** Beth and her friend Gino have a new paddling pool. Their pool holds 100 litres of water.

a) Beth and Gino fill their pool with a bucket. The bucket holds 10 litres of water. How many bucketfuls of water

will fill the paddling pool? _____ bucketfuls

b) Beth's friend Grace's pool holds twice as much as Beth's. How many litres of water does Grace's pool

hold? _____ litres

Comparing Measurements

c) How many 10 litre bucketfuls of water fill Grace's

pool? _____ bucketfuls

d) What is the total volume of water of both pools?

_____ litres

e) How many 10 litre bucketfuls is that? _____ bucketfuls

 5 marks

Marks.......... /5

Challenge 3

 1 Max grew a pumpkin for the pumpkin
competition. He used a scale to weigh
his pumpkin.

a) How many grams did Max's pumpkin

weigh?_____ g

b) Katie's pumpkin weighed 400g more than Max's. Katie's

pumpkin weighed _____ g

c) What is the weight of Katie's pumpkin in kilograms?

_____ kg

d) Tom's pumpkin weighed the same as Max's and Katie's
together! How much did Tom's pumpkin weigh?

_____ kg

e) If all three pumpkins were weighed together, what

would their total weight be? _____ kg

 5 marks

Marks.......... /5

Total marks .... /13   How am I doing?

Measuring Temperature

Challenge 1

PS **1** Each thermometer shows a different temperature. Write down each temperature.

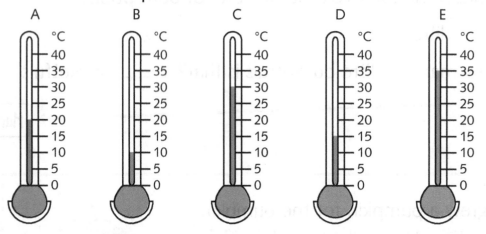

a) Thermometer A shows a temperature of _____°C.

b) The temperature of thermometer B is _____°C.

c) The third thermometer shows a temperature of _____°C.

d) Thermometer D is _____°C.

e) Thermometer E shows a temperature of _____°C.

5 marks

Marks.........../5

Challenge 2

PS **1** Colour these thermometers to show the following temperatures.

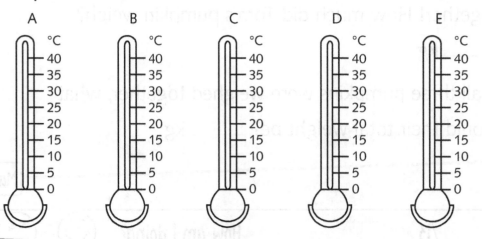

a) Thermometer A is measuring a temperature of **5°C**.

b) The temperature of thermometer B shows **40°C**.

c) The third thermometer shows a temperature of **25°C**.

d) Thermometer D shows a temperature of **15°C**.

e) Thermometer E shows **35°C**.

5 marks

Marks.......... /5

Challenge 3

PS | 1 | Start at the arrow showing 15°C.

a) What would the temperature be if it was 5°C **hotter**? _____°C

b) What if the temperature was 10°C less than 15°C? _____°C

c) The temperature gets 15°C **colder**. It is now _____°C

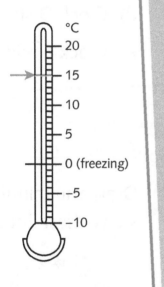

3 marks

Marks.......... /3

Total marks .... /13

How am I doing?

Measuring Time

PS Problem-solving questions

Challenge 1

1 Write the time shown by each clock.

a) Clock A shows _____.

b) Clock B shows _____.

c) Clock C tells us the time is _____.

d) Clock D shows _____.

e) Clock E tells a time of _____.

 5 marks

Marks.......... /5

Challenge 2

1 Draw the minute and hour hands on these clocks to show these times.

a) Half past 3   **b)** Quarter past 2   **c)** 10 o'clock

Measuring Time

d) Quarter to 8   **e)** Half past 10

5 marks

Marks.......... /5

Challenge 3

PS   1   Use the information in the chart to answer these questions.

60 seconds = 1 minute	7 days = 1 week
60 minutes = 1 hour	52 weeks = 1 year
24 hours = 1 day	12 months = 1 year

a) How many seconds are there in 2 minutes?

_____ seconds

b) How many hours are there in 2 days?

_____ hours

c) How many days are there in 3 weeks?

_____ days

d) How many weeks are there in 2 years?

_____ weeks

e) How many months are there in 10 years?

_____ months

5 marks

Marks.......... /5

Total marks .... /15   How am I doing?

Standard Units of Money

Challenge 1

1. How much money is there in total? _____p

☐ 1 mark

〉PS〉 2. How much money do you see this time? _____p

Write a calculation that shows how you could have worked this out. _____

☐ 2 marks

〉PS〉 3. How much money is in each purse?

A

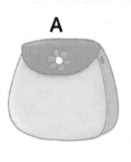

B

C

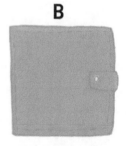

Purse A: _____p   Purse B: _____p   Purse C: _____p

☐ 3 marks

Marks.........../6

Standard Units of Money

Challenge 2

1 Find four different ways to make £10 using these pound coins and notes.

_____

_____

_____

_____

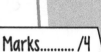
4 marks

Marks.......... /4

Challenge 3

1 Write the answers.

a) £2 + £1 = £_____

b) £5 + £1 + £2 = £_____

c) £10 + 20p + 20p = £_____

d) £5 + £5 + 10p = £_____

e) £1 + £2 + £5 + 50p = £_____

5 marks

2 Paulo has saved £7.84. Show two different ways he could have this amount using coins and notes.

_____   _____

2 marks

Marks.......... /7

Total marks .... /17   How am I doing?

Money Problems

Challenge 1

PS **1** Solve these money problems.

a) A watermelon costs 60p. An apple costs 20p. How much more does the watermelon cost? _____p

b) Adam buys 3 bananas for 90p. How much does each banana cost? _____p

c) Pears are 10p each. How much would 6 pears cost?

_____p

d) Grapes are £1.50 per bunch. Elise has £3. How many bunches can she buy? _____ bunches

e) Plums are 30p for two. How much does each plum cost? _____p

5 marks

Marks.......... /5

Challenge 2

PS **1** Joshua has been shopping and now he's ready for a day on the beach!

a) Joshua's sunglasses cost £1.50. He gave the shopkeeper £2. How much change did Joshua get? _____p

b) The bucket and spade cost £2.50 and his shoes cost £3.50. How much did Joshua spend on these items? £_____

Money Problems

c) Joshua spent £4 on his shorts and shirt. He gave the shopkeeper a £5 note. The shopkeeper gave him

£_____ change.

d) How much did Joshua spend in total? £_____

e) Joshua had £20 before he went shopping. How much money does he have left? £_____

5 marks

Marks.......... /5

Challenge 3

PS 1 Becky has been saving. She has sorted her money into coins of the same type.

a) What is the value of all the 1p coins? _____p

b) Becky has _____p worth of 2p coins.

c) The 5p coins are worth a total of _____p.

d) How much has Becky saved in total? £_____

4 marks

Marks.......... /4

Total marks .... /14   How am I doing?

2-D Shapes

Challenge 1

1 Draw lines to join the shape to its name.

Pentagon

Rectangle

Hexagon

Triangle

Circle

5 marks

PS **2** Write the names of the shapes.

a) I have five straight sides of equal length.
I am a _____.

b) I have no straight sides and no corners.
I am a _____.

c) I have four straight sides. Two of my sides are longer than the other two. I am a _____.

d) I have six straight sides. I am a _____.

e) I have three straight sides. I am a _____.

5 marks

Marks......... /10

2-D Shapes

Challenge 2

1 Draw five different rectangles in the space below.

 5 marks

Marks.......... /5

Challenge 3

1 Tick the shapes that have a correct line of symmetry.

A B C D E F

☐   ☐   ☐   ☐   ☐   ☐

3 marks

2 Draw the missing side of this shape to make it symmetrical.

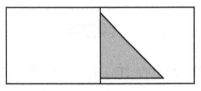

1 mark

3 Mark the corners on these shapes.

a) 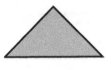   b)    c)   d)

4 marks

Marks.......... /8

Total marks .... /23   How am I doing?

81

3-D Shapes

Challenge 1

 1 Complete this table. Put a ✓ if the property is correct and a ✗ if it is not correct.

Shape	Has Six Faces	Has Circular Faces
cuboid 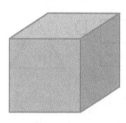		
cylinder		

 4 marks

2 **Vertices** are where two edges meet to make a corner.

a) Mark the **vertices** you can see on this cube.

b) How many edges does the cube have in total?

_____ edges

c) How many corners does the cube have in total?

_____ corners

3 marks

Marks.......... /7

Challenge 2

 1 This 3-D shape is a pyramid. Write two properties that describe it.

_____ and _____

2 marks

2 Tick the shapes that are cuboids.

A ⬚ B ⬚ C ⬚ D ⬚ E ⬚

2 marks

Marks........../4

PS **1** Write the name of the 3-D shape.

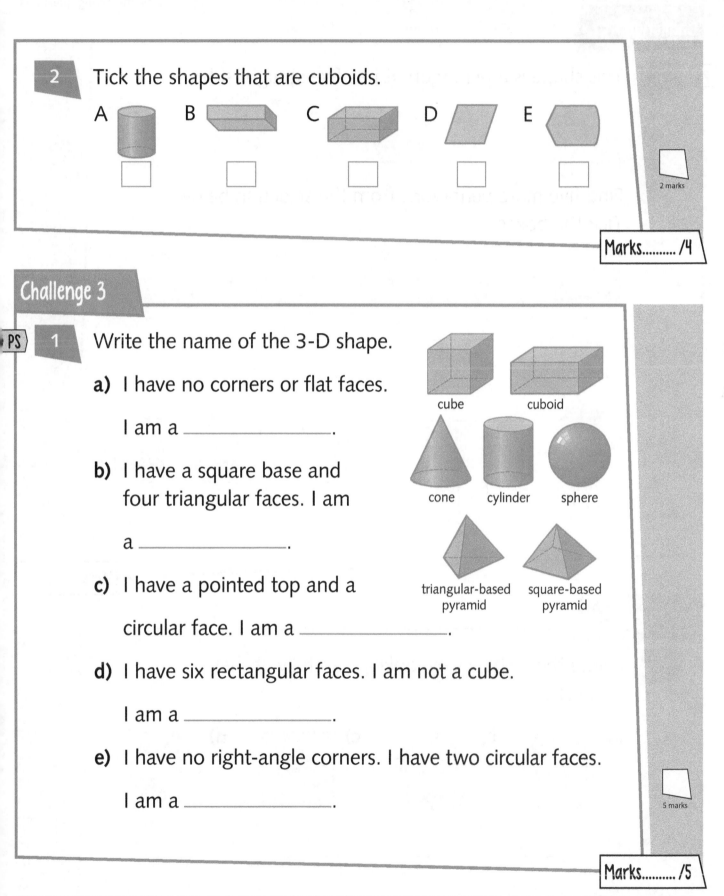

cube   cuboid

cone   cylinder   sphere

triangular-based pyramid   square-based pyramid

a) I have no corners or flat faces.

I am a _____ .

b) I have a square base and four triangular faces. I am

a _____ .

c) I have a pointed top and a

circular face. I am a _____ .

d) I have six rectangular faces. I am not a cube.

I am a _____ .

e) I have no right-angle corners. I have two circular faces.

I am a _____ .

5 marks

Marks........../5

Total marks .... /16   How am I doing?

83

Different Shapes

PS Problem-solving questions

Challenge 1

PS **1** This shape is a pentagon. It has five straight sides.

Find five more pentagons from the selection below.
Tick the boxes.

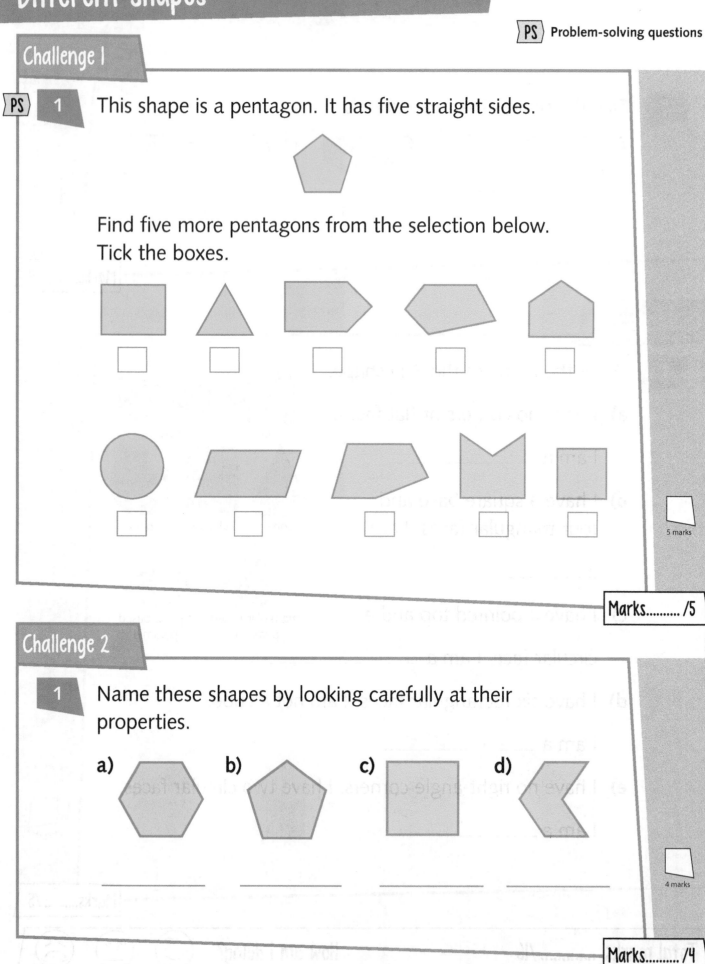

5 marks

Marks.......... /5

Challenge 2

1 Name these shapes by looking carefully at their properties.

a) b) c) d)

_____ _____ _____ _____

4 marks

Marks.......... /4

Different Shapes

Challenge 3

1 This is an example of a quadrilateral.

Draw six quadrilaterals in the space below.

6 marks

PS **2** Everyday objects have shapes like 3-D or solid shapes.

a) Name something that is a cylinder shape.

_____

b) Name something that might be cuboid in shape.

_____

c) What might you have that is spherical?

_____

d) Name something that is a cube.

_____

4 marks

Marks.........../10

Total marks ..../19

How am I doing?

Progress Test 3

PS Problem-solving questions

PS **1.** Look at the fish. There are 15 in total.

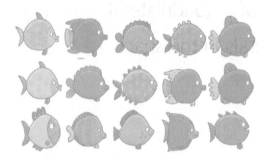

Write nine addition number facts for the total number of fish.

Example: 14 + 1 = 15

9 marks

2. Count the money.

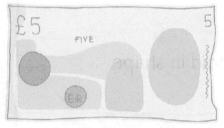   

What is the total amount? £_____

1 mark

PS **3.** **a)** If one kitten has four legs, how many legs do four kittens have? _____ legs

b) Show how you worked this out. _____

c) How many legs would ten kittens have? _____ legs

3 marks

PS **4.** Ashleigh is collecting apples.

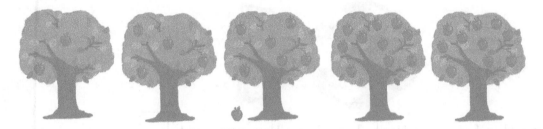

a) Count the total number of apples that she will be able to collect. Don't forget any that have already fallen off!

_____ apples

b) Ashleigh puts ten apples in each basket. How many baskets will she need?_____ baskets

c) Last year Ashleigh collected double the number of apples. How many did she collect last year? _____ apples

3 marks

PS **5.** Answer these capacity questions.

a) If the capacity of one jug is 200 ml, what is the capacity of four jugs? _____ ml

b) A bath holds 100 litres of water. If it fills at 10 litres per minute, how long, in minutes, will it take? _____ minutes

c) David needs to fill his 20 litre fish tank. He has a 200 ml container. How many containers will he use to fill the tank?

_____ containers

d) Jason has 200 ml of ice cream. How many 50 ml scoops can he serve? _____ scoops

4 marks

6. Do these fraction problems.

    a) Write the fraction that describes the amount of footballs.

    b) If you added $3\frac{1}{2}$ footballs to them, how many footballs would you have? _____ footballs

2 marks

7. Put the correct symbol <, > or = in the gaps.

    a) $\frac{1}{2}$ of 20 ☐ $\frac{1}{4}$ of 40

    b) $\frac{3}{4}$ of 80 ☐ $\frac{1}{2}$ of 100

    c) $\frac{2}{4}$ of 24 ☐ $\frac{1}{2}$ of 24

    d) $\frac{3}{4}$ of 40 ☐ $\frac{1}{2}$ of 60

4 marks

PS **8.** Answer these questions about length.

    a) The snake measures exactly 50 cm. What would half a snake measure? _____ cm

    b) What would be the total length of four snakes, in metres? _____ m

    c) How many snakes would be needed to measure 3 m? _____ snakes

3 marks

9. Colour these thermometers to show the following temperatures.

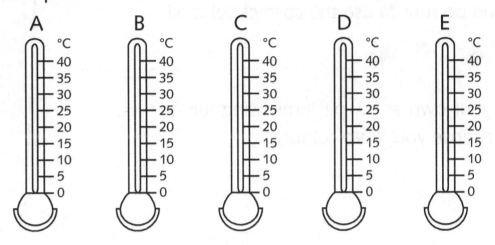

A   B   C   D   E

a) Thermometer A shows 15°C.

b) Thermometer B shows a temperature twice as hot as A.

c) Thermometer C is 5°C colder than A.

d) Thermometer D shows 20°C.

e) Thermometer E shows 35°C.

5 marks

PS **10.** Answer these real-life money problems.

a) Kane has 60p. He spends 50p on his favourite magazine. How much does he have left? _____p

b) Maisie visits the pet shop and spends £1.50. She gives the shopkeeper a £2 coin. How much change does she receive? _____p

c) Shamir saves 50p each week. How much does he save in 4 weeks? £_____

3 marks

Marks........ /37

Patterns

Challenge 1

PS | 1 | Continue this arrow pattern by drawing the next six arrows and be sure to use the correct colours!

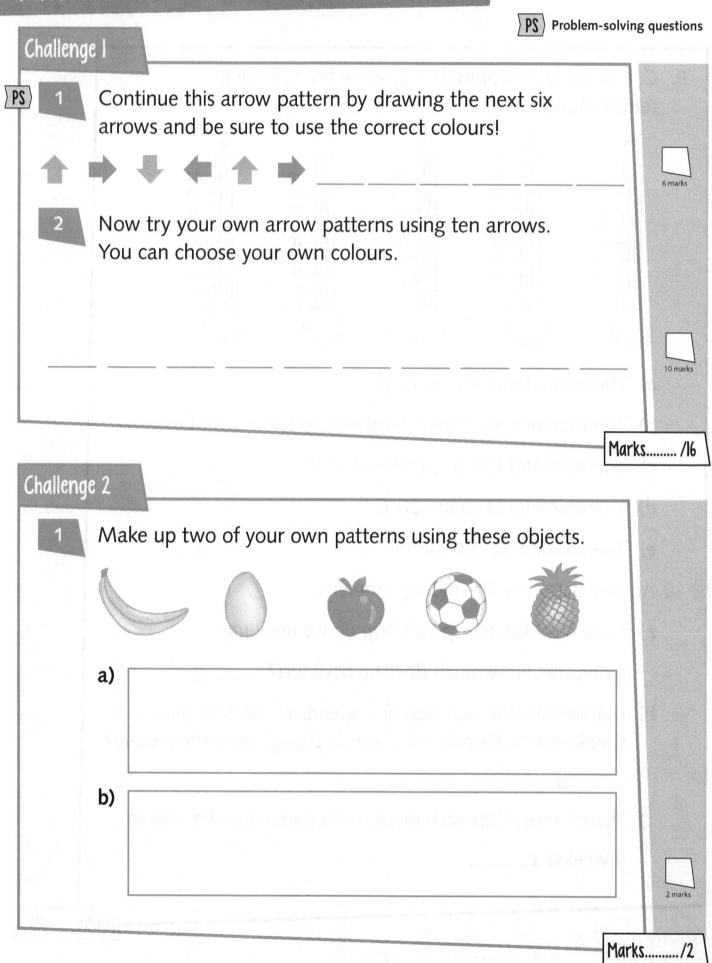

6 marks

2 | Now try your own arrow patterns using ten arrows. You can choose your own colours.

10 marks

Marks......... /16

Challenge 2

1 | Make up two of your own patterns using these objects.

a)

b)

2 marks

Marks.......... /2

90

Challenge 3

1   Use your own ideas to make two new patterns.

a)

b)

10 marks

Marks......... /10

Total marks .... /28   How am I doing?

Sequences

Challenge 1

1 Continue these number sequences.

a)

20	20	19	19						

b)

1,1,1	3,3,3	5,5,5			

c)

5,5,10	6,6,12			

d)

100	100	90	90					

4 marks

Marks.........../4

Challenge 2

1 Fill in the missing parts of these sequences.

a)

b) 1, 1, 1, 3, 2, 2, 2, 6, ____, ____, ____, 9, 4, 4, ____, ____

c) Now try this really tricky one!

20	10	30	20	20	40	20					

3 marks

Marks.........../3

92

Challenge 3

1 Use the following numbers and shapes to make a pattern.

> **Example:** If your numbers were 1, 2, 2 you could use 1 square, 2 triangles, 2 circles, 1 square, 2 triangles . . .

You must use these numbers in this order.

3	1	2	3	3	1	2	3

Choose any of these shapes for your pattern.

Now create your pattern.

5 marks

2 Choose a different sequence of numbers and your own shapes to make a new pattern.

5 marks

Marks........ /10

Total marks .... /17   How am I doing?

93

Quarter Turns and Half Turns

PS Problem-solving questions

Challenge 1

PS **1** Charlie is lost; he needs help to find his way around!

Right ➡

⬅ Left

	X	
Z	Charlie	W
	Y	

a) If Charlie turned to the left, which letter would he see?

_____

b) Charlie starts again, turns right and sees the letter _____.

c) Which letter is above Charlie? _____

d) Which letter does Charlie have below him? _____

4 marks

Marks.......... /4

Challenge 2

PS **1** Take a look at the circle.

a) If you stood on the **X** at the centre, as Charlie started above, would you be facing a letter? Yes/No _____

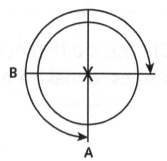

b) Make a half turn to the **left**. Which letter do you see? _____

c) Start facing **A** and make a quarter turn to the right. Which letter do you see? _____

Quarter Turns and Half Turns

d) Start facing **B** and make a $\frac{3}{4}$ turn to the right. Do you see a letter? What is it? _____

e) Face **A** and make a half turn to the left. Can you see a letter? Yes/No _____

5 marks

Marks.......... /5

Challenge 3

PS **1** Help Charlie's brother Sam find his pets.

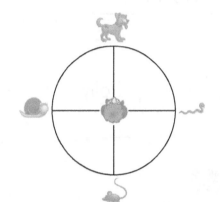

a) Which pet is Sam facing?

_____

b) If Sam turns half a turn to the right, which pet can he now see? _____

c) Sam is now facing the snail. He makes a quarter turn to the left. What does he see? _____

d) Sam is facing the snake and makes a three quarter turn to the right, then a quarter turn left. Which pet does he see? _____

e) If Sam faces the mouse and makes a half turn to the left and then a quarter turn to the right, what can he see?

_____

5 marks

Marks.......... /5

Total marks .... /14   How am I doing?

Right-Angle Turns

Challenge 1

 PS **1** Help Mia find her favourite lunch.

	pizza	
ice cream	Mia	banana
	lolly	

a) Mia is facing the pizza. If she makes a right-angle clockwise turn, what can she see?

_____

b) Mia is now facing the lolly. She makes a half turn anticlockwise. What is she facing now?

_____

2 marks

Marks........../2

Challenge 2

 PS **1** Follow the instructions to complete the grid.

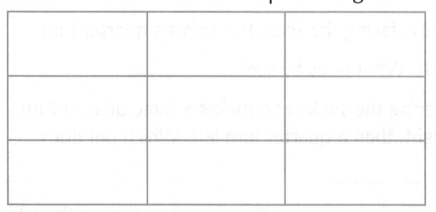

a) Draw the top of someone's head in the centre square facing up the page.

b) Draw something to eat in the square that would be a clockwise right-angle turn for the face.

c) Draw your favourite toy above the face.

d) Draw an alien in the square that is a quarter anticlockwise turn for the face.

e) Draw your ideal pet below the face.

5 marks

Marks.......... /5

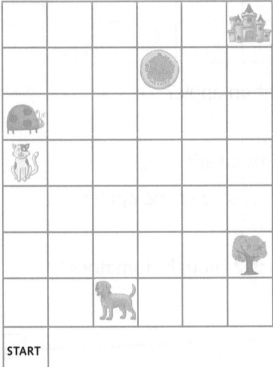

START

a) Begin at START. Go forward four squares. What have you found?

_____

b) Now make a right-angle **clockwise** turn. Then move on two squares. Make another right-angle **clockwise** turn. Then move on three squares. What have you found?

_____

c) Remember which way you are facing! From here make a right-angle **anticlockwise** turn and move on three squares. Make another right-angle **anticlockwise** turn and move forward one square. What do you have?

_____

d) Move forward five squares and you are at the

_____ .

4 marks

Marks.......... /4

Total marks .... /11   How am I doing?

97

Pictograms

Challenge 1

PS | 1 | Class 2 have made a pictogram of their healthy snacks. Each person brought one snack.

Apple	🍎	🍎	🍎			
Tomato	🍅	🍅	🍅	🍅	🍅	
Carrot	🥕	🥕				
Banana	🍌	🍌	🍌	🍌	🍌	🍌

a) How many children brought an apple?

_____ children

b) Which snack was the least popular? _____

c) Which snack had double the score of the apples?

_____

d) How many members of class 2 brought tomatoes?

_____ members

4 marks

Marks.......... /4

Challenge 2

PS | 1 | Scott and his sister made a pictogram to show how much traffic passed their house in 10 minutes. There was so much traffic they had to make each picture worth **two** vehicles!

Car	🚗	🚗	🚗	🚗	🚗	🚗
Van	🚐	🚐	🚐			
Bus	🚌	🚌				
Motorbike	🏍	🏍	🏍			

a) How many buses passed by? (Remember what each

picture is worth). _____ buses

b) Most of the vehicles that passed were _____.

There were _____ of them.

c) How many motorbikes passed? _____ motorbikes

d) What was the total number of vans? _____ vans

Marks........... /4

Challenge 3

PS **1** This pictogram shows the snails that Henry found in his garden during the summer holidays.

a) How many type C snails did

Henry find? _____ snails

b) _____ type B snails were discovered.

c) Henry found the same number

of which three snails? _____,

_____ and _____.

d) How many type E snails did Henry find in his garden?

_____ snails

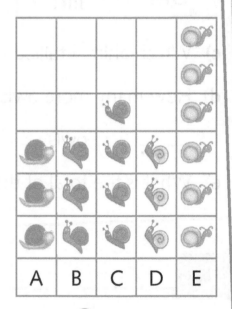

 = two snails

4 marks

Marks........... /4

Total marks .... /12   How am I doing?   😖

Tally Charts

Challenge 1

PS 1 Year 2 went to a wildlife park. They did a tally of the animals they saw.

Animal	Tally	Total
Meerkat	卌 卌 I	
Goat	卌 III	
Duck	卌 卌 卌	
Otter	IIII	
Owl	卌 I	

a) Complete the table by filling in the totals column.

b) What were most of the animals Year 2 saw?

_____

c) Which animal tallied the fewest? _____

d) How many more owls did they see than otters?

_____ owls

e) What is the total of the complete tally? _____

5 marks

Marks.......... /5

Challenge 2

1 Class 3 made a tally of hair colours in their class.

Hair Colour	Tally	Total
Blond		7
Black		6
Red		5
Brown		11

a) Show the totals as a tally.

100

b) How many more children had brown hair than black?

_____ children

c) Which two hair colours equal the brown total?

_____ and _____

d) How many children took part in the tally?

_____ children

4 marks

Marks.......... /4

Challenge 3

PS   1   Amir has collected groups of fruit and vegetables.

a) Count the different fruits and vegetables and make a tally of each one in the tally column.

Food	Tally	Total
Banana		
Potato		
Apple		
Tomato		

b) Complete the totals column.

c) How many items did Amir have in total? _____ items

3 marks

Marks.......... /3

Total marks .... /12   How am I doing?

Block Diagrams

 PS Problem-solving questions

Challenge 1

 1 Wendy and her friends have been counting the number of teeth that they have put under their pillows over the last year. They have made a block diagram to show the data.

a) How many teeth has Wendy placed under her pillow?

_____ teeth

b) Kate has put _____ teeth under her pillow.

c) How many more teeth has Holly put under her pillow

than Bella? _____ teeth

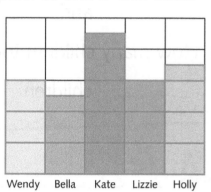

Wendy   Bella   Kate   Lizzie   Holly

= two teeth

d) Which of Wendy's friends has lost the fewest teeth?

_____

e) How many teeth have Kate and Holly lost altogether?

_____ teeth

5 marks

Marks.......... /5

Challenge 2

1 The following block diagram shows data collected by camera traps in the school grounds.

Look carefully at the diagram and **the steps it is counting in** to answer these questions.

a) How many photographs of a badger were there?

_____ photographs

b) There were _____ mouse photographs.

c) Which animal had the same number of photographs as foxes and badgers combined?

_____

d) How many more mouse photographs were there than fox photographs?

_____ photographs

e) What was the total number of animal photographs?

_____ photographs

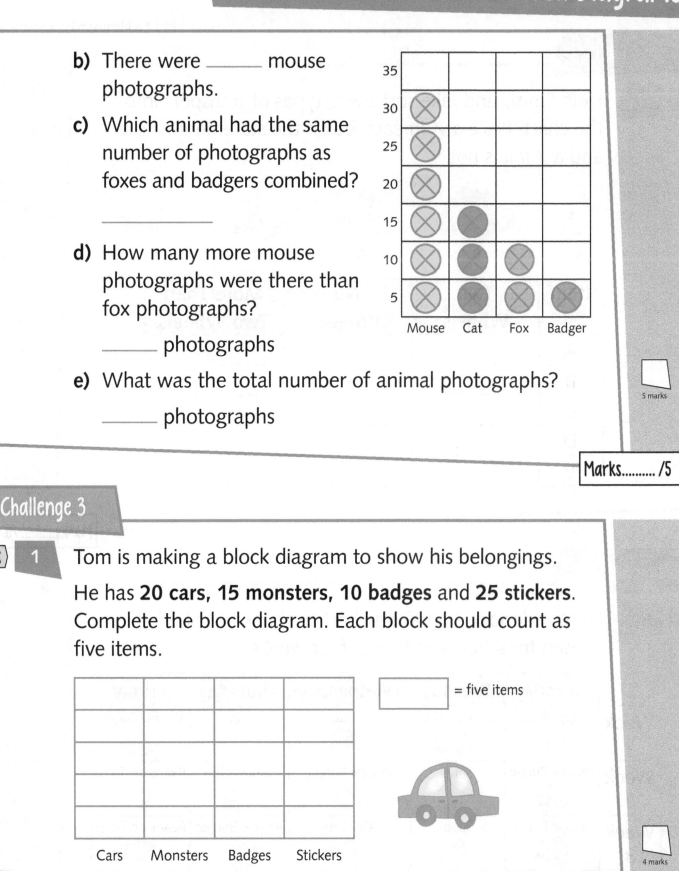

5 marks

Marks.......... /5

Challenge 3

PS 1 Tom is making a block diagram to show his belongings.

He has **20 cars, 15 monsters, 10 badges** and **25 stickers**. Complete the block diagram. Each block should count as five items.

☐ = five items

Cars   Monsters   Badges   Stickers

4 marks

Marks.......... /4

Total marks .... /14   How am I doing?

Tables

Challenge 1

PS 1 Help Emmy and Jake sort these types of transport into the chart. Place a ✓ in each box if the statement is true and a ✗ if it is not.

A   B   C   D   E

Item	Two Wheels	No Wheels	More Than Two Wheels
A			
B			
C			
D			
E			

5 marks

Marks.......... /5

Challenge 2

PS 1 Frankie decided to make a chart to show what he had chosen for school lunch over four weeks.

	Monday	Tuesday	Wednesday	Thursday	Friday
Week 1	Veggie Burger	Sandwiches	Salad	Soup	Veggie Burger
Week 2	Veggie Burger	Soup	Veggie Burger	Sandwiches	Beans on Toast
Week 3	Veggie Burger	Soup	Chicken	Veggie Burger	Beans on Toast
Week 4	Salad	Veggie Burger	Salad	Chicken	Sandwiches

a) Which lunch did Frankie eat most often? _____

How many times did he have it? _____ times

b) How many times did Frankie order soup? _____ times.

c) Frankie chose salad the same number of times as which

other two meals? _____ and _____

d) Which other meals did Frankie only order twice?

_____ and _____

4 marks

Marks.......... /4

Challenge 3

1 Take a look at these animals to complete this chart. Place
a ✓ or ✗ in each box.

A   B   C   D   E   F

Animal	Long Neck	Four Legs	Spots or Stripes
A			
B			
C			
D			
E			
F			

6 marks

Marks.......... /6

Total marks .... /15   How am I doing?

105

Gathering Information and Using Data

Challenge 1

PS **1** Tim is about to have his biggest ever toy sale. Make a tally and total of each type of toy.

Toy	Tally	Total

10 marks

Marks......... /10

Challenge 2

PS **1** Use the data on the tally chart to answer these questions.

a) Which toy does Tim have most of? _____

How many does he have? _____

b) How many robots does Tim have? _____ robots

This is equal to the number of _____.

Gathering Information and Using Data

c) How many toys are there in total? _____ toys

d) How many fewer aliens are there than robots?

_____ aliens

e) When you add together the aliens, footballs and

teddies, what is the total? _____

7 marks

Marks.......... /7

Challenge 3

PS 1 Use the data to create a block diagram.

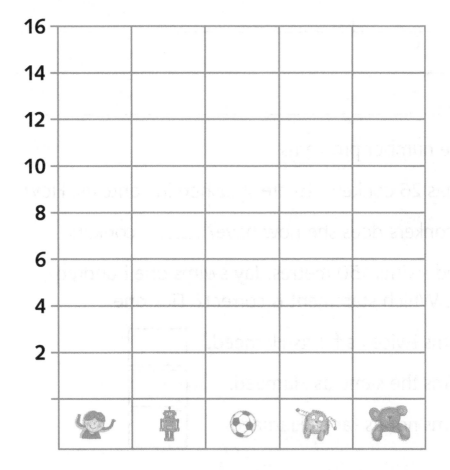

5 marks

Marks.......... /5

Total marks .... /22   How am I doing?

Progress Test 4

PS 1. Put these numbers in order from least value to most.

23 3 17 21 79 97 85 58

[ ] [ ] [ ] [ ] [ ] [ ] [ ] [ ]

1 mark

2. Write the addition and subtraction families for these sets of numbers.

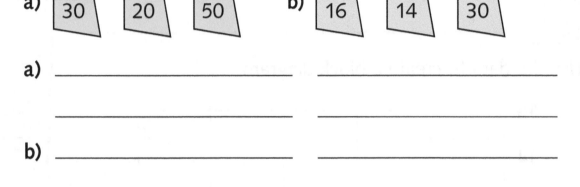

a) 30 20 50   **b)** 16 14 30

a) _____   _____

_____   _____

b) _____   _____

_____   _____

4 marks

PS 3. Solve these number problems.

a) Anna has 26 conkers. 16 are smashed in contests. How many conkers does she now have? _____ conkers.

b) Hameed swims 150 metres. Jay swims one hundred metres. Which statement is correct? Tick **one**.

Jay swims twice as far as Hameed. [ ]

Jay swims the same as Hameed. [ ]

Jay swims not as far as Hameed. [ ]

c) Add seven, twenty-three and five. _____

d) Subtract 12 from 37. _____

4 marks

4. Here are the different leaves that Allen found on a visit to the park. Use the chart to help Allen sort them out. Place a ✓ in each box if the statement is true and a ✗ if it is not.

A   B   C   D   E

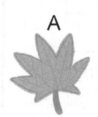   

Leaf	Has Fewer Than 3 Points	Has More Than 3 Points	Smooth Edges	Jagged Edges
A				
B				
C				
D				
E				

5 marks

5. Answer these divisions to halve the numbers.

> **Example:** 14 ÷ 2 = 7   Half of 14 is 7.

a) 22 ÷ 2 = _____   Half of 22 is _____.

b) 18 ÷ 2 = _____   Half of 18 is _____.

c) 30 ÷ 2 = _____   Half of 30 is _____.

d) 100 ÷ 2 = _____   Half of 100 is _____.

e) 60 ÷ 2 = _____   Half of 60 is _____.

5 marks

6. a) Draw lines to join the objects to their estimated weight.

A Sheet of paper

About 1 kg

B Comic book

More than 5 kg

C Bowling ball

Less than 5 g

D Packet of flour

About 50 g

5 marks

b) Which object is the lightest? _____

7. Write the answers and multiplications for these repeated additions.

a) 10 + 10 + 10 = _____   **b)** 5 + 5 + 5 + 5 + 5 = _____

_____ × _____ = _____   _____ × _____ = _____

2 marks

 8. a) This aeroplane takes two hours to fly from London to Spain. If it departs at 2 o'clock, what time will it land? _____

b) How many hours would it take for the plane to fly to Spain and back to London? _____ hours

2 marks

9. How many quarters do these equal?

a) $3\frac{1}{4}$ _____ quarters   **b)** $2\frac{3}{4}$ _____ quarters

c) $1\frac{1}{2}$ _____ quarters   **d)** 5 _____ quarters

4 marks

10. Follow the instructions below to complete the grid.

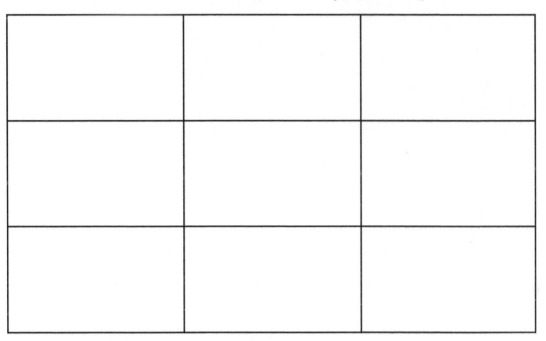

   a) Draw the top of someone's head in the centre box facing up the page.

   b) Draw a diamond in the box that would be an anticlockwise right-angle turn for the face.

   c) Draw flowers below the face.

   d) Draw a cat in the box that would be a quarter clockwise turn for the face.

   e) Draw a ghost in the box above the face.

5 marks

11. Draw a line of symmetry on each shape.

   a) 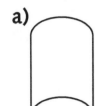   b) 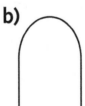   c)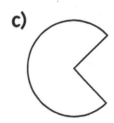

3 marks

Marks........ /40

Pages 4–11
Starter Test

1. a) 3, **4**, 5, 6, **7**, **8**, **9**, 10, **11**, 12
   b) 20, **19**, 18, **17**, 16, **15**, **14**, 13, **12**, 11
   c) 2, **4**, 6, 8, 10, 12, **14**, **16**, **18**, **20**

2. a)

6	12	16	31	87

   b)

0	2	15	30	60

   c)

14	15	23	99	100

3. a) Any two segments shaded.
      E.g.

   b) Any two segments shaded.
      E.g.

   c) Line splits shape vertically and
      1 side shaded.
      E.g.

4. a) three   b) ten   c) twenty
5. a) 7   b) 12   c) 9
6. a) Truck D   b) Truck C   c) Shorter
   d)

D	A	B	C

   e) 26 cm

7. a)

14	15	**16**

   b)

19	20	**21**

   c)

44	45	**46**

8. a)

45	50	**55**

   b)

30	35	**40**

   c)

15	20	**25**

9. a)

   b)

   c)

10. a) 6 + 2 = 8 **OR** 2 + 6 = 8, 8 − 6 = 2 **OR**
       8 − 2 = 6
    b) 9 + 1 = 10 **OR** 1 + 9 = 10, 10 − 9 = 1 **OR**
       10 − 1 = 9
    c) 7 + 5 = 12 **OR** 5 + 7 = 12, 12 − 7 = 5 **OR**
       12 − 5 = 7

11. a) 16   b) 4
12. a) $\frac{1}{4}$ b) $\frac{3}{4}$
13. 0, 5, **10**, 15, 20, **25**, **30**, **35**, 40, **45**, 50
14. a) >   b) <   c) =
15. a) 6   b) Snails
    c) 4   d) 16
16. 500 g
17. a) 20   b) 25
    c) 30   d) 14
    e) 40   f) 60
18. 31p
19. a) 6   b) 3   c) 10
20. a) $\frac{1}{2}$ b) $\frac{1}{4}$
21. Any six of following:
    10 + 0 = 10, 0 + 10 = 10, 1 + 9 = 10,
    9 + 1 = 10, 2 + 8 = 10, 8 + 2 = 10, 3 + 7 = 10,
    7 + 3 = 10, 6 + 4 = 10, 4 + 6 = 10, 5 + 5 = 10
22. a) triangle
    b) square
    c) rectangle or oblong
23.
24. a) 5   b) 6
    c) 10   d) 4
25. a) 6   b) 10
    c) 20   d) 12
26.

Pages 12–13
Challenge 1

1. a) 11
   b) 11, 21, 33, 45, 56
2. a) 71
   b) 14, 18, 22, 50, 71
3. a) **20**, 21, **22**, **23**, 24, 25, **26**, 27, **28**, **29**
   b) 30, 31, **32**, **33**, **34**, 35, **36**, 37, **38**, 39
4. a) Any number from 51 to 59
   b) Any number from 46 to 60

Answers

Challenge 2
1 a) 73 b) 35
   c) 87 d) 56
   e) 17 f) 91
   g) 59 h) 12
   i) 22 j) 106

Challenge 3
1 a) Eighteen
   b) Ninety-seven
   c) Fifty-six
   d) Fifteen
   e) Eighty-one

Pages 14–15
Challenge 1
1 a) 21 b) 14
   c) 18 d) 11
   e) 17 f) 16
2 a) 26 b) 23
   c) 19 d) 27
   e) 29 f) 15

Challenge 2
1 a) 21
   b) 27
   c) 42
   d) 42
   e) 41

2

1	**2**	3	**4**	5	**6**	7	**8**	9	**10**
11	**12**	13	**14**	15	**16**	17	**18**	19	**20**

Challenge 3
1 13, 23, 34, 45, 62, 76, 89
2 a) 38 b) 30
   c) 31 d) 35

Pages 16–17
Challenge 1
1 a) 16, 18, 20, 22, **24**, **26**, **28**, **30**
   b) 23, 21, 19, 17, **15**, **13**, **11**, **9**
   c) 92, **94**, 96, 98, **100**, 102, **104**
2 a) 2
   b) 10

Challenge 2
1 a) 45, **50**, 55, 60, **65**, 70, **75**
   b) 100, **95**, 90, **85**, 80, **75**, 70
   c) **26**, 31, 36, **41**, 46, **51**, 56
2 a) 41, 36, 31
   b) 54, 59, 64
   c) 79, 74, 69

Challenge 3
1

a)	b)	c)	d)
58	**65**	**51**	37
53	**70**	**46**	32
48	75	**41**	27
43	**80**	**36**	22
38	**85**	**31**	17
33	90	26	12
28	**95**	21	**7**
23	100	16	**2**

2 a) 3   b) 5   c) 3

Pages 18–19
Challenge 1
1 a) 50
   b) 4
   c)

30	40	50	60	70	80	90	100

   d) 90
   e) 30
   f) 50
   g) 4

Challenge 2
1 5, 15, 25, 35, 45, 55, 65, 75
2 80, 70, 60, 50, 40, 30, 20, 10
3 a) 15, 18, 21, 24
   b) 12, 9, 6, 3
   c) 2

Challenge 3
1

6	**9**	12	**15**	18
21	**31**	**41**	51	**61**
14	16	**18**	20	**22**
67	**64**	61	**58**	55

Pages 20–21
Challenge 1
1

64	65	66	**77**	78	**79**	**18**	19	20
98	99	**100**	**119**	120	**121**	**46**	47	**48**

2 a) 31 b) 39 c) 70
   d) 100 e) 120 f) 93

Challenge 2
1 a) 23 b) 81
   c) 77 d) 13
2 a) 56 b) 88
   c) 101 d) 17
3 42

Challenge 3
1. a) 34   b) 66   c) 100
   d) 142   e) 184   f) 233
2. a) 23   b) 4
3. a) 15   b) 27   c) 62

Pages 22–23
Challenge 1
1. a) 63   b) 19
   c) 55   d) 83
2. a)

   b)
   c)
   d)

Challenge 2
1. a) 80 + 2   b) 10 + 5
   c) 90 + 8   d) 30 + 6
2. a) 65   b) 51
   c) 72   d) 47
   e) 19   f) 36

Challenge 3
1. Any number from 22–33
2. Any number from 81–99
3. a) 3 tens, 4 ones
   b) 8 tens, 4 ones
   c) 6 tens, 6 ones
4. a)

   1   7   12   21
   b)

   27   36   45   98

Pages 24–25
Challenge 1
1. Appropriate answers. Examples below:
   a) 10 is less than 12
   b) 54 is greater than 23
   c) 23 is equal to 23
   d) 66 is greater than 54

Challenge 2
1. a) <   b) >   c) <
2. a) <   b) >   c) =

Challenge 3
1.

15	=	15
Appropriate number	>	Appropriate number
54	<	84
Appropriate number	=	Appropriate number
17	>	15

Award 1 mark for each correct row.
2. a) equal to
   b) 4 more
   c) 3 times more than / 6 more than
   d) bananas, oranges, apples

Pages 26–27
Challenge 1
1. a) Double 10 is 20, half of 20 is 10.
   b) Double 15 is 30, half of 30 is 15.
   c) Double 100 is 200, half of 200 is 100.
2. a) Double 12 is 24, half of 24 is 12.
   b) Double 20 is 40, half of 40 is 20.
   c) Double 200 is 400, half of 400 is 200.

Challenge 2
1. a) 20   5 + 15 = 20
   b) 33   10 + 23 = 33
   c) 46   12 + 34 = 46
   d) 68   14 + 54 = 68
2. a) 15 + 20 = 35
      20 + 15 = 35
      35 − 15 = 20
      35 − 20 = 15
   b) 45 + 15 = 60
      15 + 45 = 60
      60 − 45 = 15
      60 − 15 = 45
   c) 30 + 70 = 100
      70 + 30 = 100
      100 − 30 = 70
      100 − 70 = 30
   d) 98 + 1 = 99
      1 + 98 = 99
      99 − 98 = 1
      99 − 1 = 98

Challenge 3
1. a) 91   b) 82
   c) 73   d) 98
2. a) 34   b) 53
   c) 33   d) 61

Pages 28–29
Challenge 1
1. a) 13   b) 15
   c) 22   d) 23

Answers

Challenge 2
1 a) $15 - 6 = 9$
    b) $12 + 11 = 23$
    c) $20 - 10 = 10$
    d) $26 - 9 = 17$

Challenge 3
1   60 $20 + 20 + 20 = 60$
2   7 $22 - 15 = 7$
3   59

Pages 30–31
Challenge 1
1   Any nine appropriate number facts for 20.
Challenge 2
1. Any ten two-digit numbers using any two of
    the following: 2, 3, 4, 5, 6, 7, 8 and 9.
2   a) 33   b) 64
    c) 46   d) 26
    e) 88   f) 57

Challenge 3
1   Any nine appropriate subtractions.
2   26   31   36   41   46   51   56

Pages 32–33
Challenge 1
1   a) 8   b) 8
    c) 22   d) 10
    e) 4   f) 7
    g) 12   h) 11

Challenge 2
1   a) 13p   b) 20p
    c) 22p   d) 33p
    e) 5p

Challenge 3
1   a) +   b) –
    c) +   d) –
    e) +

2

25	26	27	28	29
35	36	37	38	39
45	46	47	48	49
55	56	57	58	59
65	66	67	68	69

Award 1 mark for each correct row.

Pages 34–37
Progress Test 1
1

1	8	3
8	3	1
3	1	8

Award 1 mark for each correct row.

2   Any appropriate two-digit number using any
    two of the following: 1, 4, 9, 6 and 3.
3   a) 13
    b) 3
    c) 18
4   a) $10 + 27$
    b) $87 + 32$
    c) $34 + 164$
    d) $120 + 76$
5   ⑧⓪
6   a) 27   b) 94
    c) 16   d) 39
    e) 14   f) 76
7   a) 32   b) 48
    c) 11
8   a) 12   b) 5
    c) 122   d) 71
    e) 33   f) 96
9   a) fifty
    b) seventeen
    c) eighty-nine
    d) thirty-one
    e) seventy-three
10 a) 50   b) 68
    c) 100   d) 26
    e) 54   f) 84
11 a) 6   b) 12
    c) 50   d) 16
    e) 8   f) 39
12 a) 12   b) 98
    c) 987   d) 123
13 a) $25 + 30 = 55$
       $30 + 25 = 55$
       $55 - 25 = 30$
       $55 - 30 = 25$
    b) $140 + 60 = 200$
       $60 + 140 = 200$
       $200 - 60 = 140$
       $200 - 140 = 60$
    c) $31 + 9 = 40$
       $9 + 31 = 40$
       $40 - 9 = 31$
       $40 - 31 = 9$
    d) $38 + 52 = 90$
       $52 + 38 = 90$
       $90 - 38 = 52$
       $90 - 52 = 38$
14 a) 72   b) 95
    c) 122   d) 71
    e) 33
15 23
16 a) 20   b) 10

Heading

Pages 38–39
Challenge 1
1 a) 25
   b) 5 + 5 + 5 + 5 + 5 = 25
   c) 5 × 5 = 25
2 a) 2 × 3 = 6
   b) 5 × 2 = 10
   c) 3 × 4 = 12
3 10p + 10p + 10p + 10p + 10p + 10p + 10p + 10p + 10p + 10p = £1

Challenge 2
1 2 × 5 = 10 and 5 × 2 = 10
2 a) Shows array 3 × 4
   b) Shows array 4 × 2
   c) Shows array 2 × 5
   d) Shows array 3 × 10

Challenge 3

Repeated Addition	Multiplication 1	Multiplication 2
3 + 3 + 3 + 3 = 12	3 × 4 = 12	4 × 3 = 12
e.g. **3 + 3 + 3 + 3 + 3 = 15**	5 × 3 = 15	**3 × 5 = 15**
10 + 10 + 10 = 30	**10 × 3 = 30**	3 × 10 = 30
e.g. **5 + 5 + 5 + 5 = 20**	**5 × 4 = 20**	4 × 5 = 20
e.g. **5 + 5 + 5 + 5 + 5 + 5 + 5 = 35**	5 × 7 = 35	**7 × 5 = 35**
2 + 2 + 2 + 2 + 2 = 10	**2 × 5 = 10**	**5 × 2 = 10**

Pages 40–41
Challenge 1
1 a) 5 badges
   b) 2 badges
   c) 25 badges

Challenge 2
1 a) 6 buns
   b) 12 ÷ 2 = 6
2 a) 4 sweets
   b) 16 ÷ 4 = 4
3 a) 5 slices   b) 15 ÷ 3 = 5

Challenge 3

Number of Items	Number of People	Division Sentence
20	4	20 ÷ 4 = 5
10	2	**10 ÷ 2 = 5**
12	**3**	12 ÷ 3 = 4
20	4	**20 ÷ 4 = 5**
18	3	**18 ÷ 3 = 6**
40	**2**	40 ÷ 2 = 20

Pages 42–43
Challenge 1

1 × 2 = 2	1 × 5 = 5	1 × 10 = 10
2 × 2 = 4	**2 × 5 = 10**	**2 × 10 = 20**
3 × 2 = 6	3 × 5 = 15	3 × 10 = 30
4 × 2 = 8	**4 × 5 = 20**	**4 × 10 = 40**
5 × 2 = 10	**5 × 5 = 25**	5 × 10 = 50
6 × 2 = 12	6 × 5 = 30	**6 × 10 = 60**
7 × 2 = 14	**7 × 5 = 35**	7 × 10 = 70
8 × 2 = 16	8 × 5 = 40	**8 × 10 = 80**
9 × 2 = 18	**9 × 5 = 45**	9 × 10 = 90
10 × 2 = 20	**10 × 5 = 50**	**10 × 10 = 100**

Challenge 2
1

1	2	3	4	5
6	7	8	9	10
11	12	13	14	15
16	17	18	19	20
21	22	23	24	25

Award 1 mark for each correct column.
2 a) odd
   b) even
   c) odd
3 a) True
   b) False

Challenge 3
1 a) 5 × 3 = 15
   b) 10 × 2 = 20
   c) 5 × 10 = 50
   d) 2 × 9 = 18
   e) 10 × 3 = 30
   f) 10 × 5 = 50

Pages 44–45
Challenge 1
1 6 ÷ 2 = 3
2 a) 15 ÷ 5 = 3 or 15 ÷ 3 = 5
   b) 20 ÷ 10 = 2 or 20 ÷ 2 = 10
3 a) 25 ÷ 5 = 5
   b) 40 ÷ 8 = 5 or 40 ÷ 5 = 8

Challenge 2
1 a) 15   30 ÷ 2 = 15
   b) 6   30 ÷ 5 = 6
   c) 3   30 ÷ 10 = 3

Challenge 3
1 a) 6 coins
   b) 3 coins
2 a) 5
   b) 6
   c) 8

Answers

d) 12

3

20	÷	5	=	**4**
14	÷	**2**	=	7
100	÷	10	=	10

Pages 46–47
Challenge 1
1   Appropriate answers, e.g.
   a) 12 ÷ 6 = 2
   b) 50 ÷ 10 = 5
   c) 15 ÷ 5 = 3
   d) 20 ÷ 10 = 2
   e) 100 ÷ 10 = 10
2   Appropriate answers, e.g.
   a) 3 × 5 = 15
   b) 7 × 5 = 35
   c) 4 × 5 = 20
   d) 2 × 10 = 20
   e) 9 × 10 = 90
Challenge 2
1   **a)** 7 × 5 = 35
      5 × 7 = 35
      35 ÷ 7 = 5
      35 ÷ 5 = 7
   b) 2 × 10 = 20
      10 × 2 = 20
      20 ÷ 10 = 2
      20 ÷ 2 = 10
2   **a)** 5, 9, 45
   b) 10, 6, 60
Challenge 3
1   **a)** 6   12 ÷ 2 = 6
   b) 24   12 × 2 = 24

Pages 48–49
Challenge 1
1   **a)** 10   10
   b) 20   20
   c) 6   6
   d) 24   24
2   **a)** 6   6
   b) 8   8
   c) 10   10
   d) 7   7
Challenge 2
1   **a)** 10
   b) 5   10 ÷ 2 = 5
   c) 20   2 × 10 = 20 or 10 × 2 = 20
2   **a)** 12
   b) 6   12 ÷ 2 = 6
   c) 24   12 × 2 = 24

Challenge 3

6 × 2 = 12	**2 × 6 = 12**	12 ÷ 2 = 6	**12 ÷ 6 = 2**
7 × 2 = 14	2 × 7 = 14	**14 ÷ 2 = 7**	**14 ÷ 7 = 2**
8 × 2 = 16	**2 × 8 = 16**	16 ÷ 2 = 8	**16 ÷ 8 = 2**
10 × 2 = 20	**2 × 10 = 20**	**20 ÷ 2 = 10**	20 ÷ 10 = 2
50 × 2 = 100	**2 × 50 = 100**	**100 ÷ 2 = 50**	**100 ÷ 50 = 2**

Pages 50–51
Challenge 1
1   **a)** 10 eggs   5 × 2 = 10
   b) 11 sheep   22 ÷ 2 = 11
Challenge 2
1   **a)** 20
   b) 10
   c) 20 ÷ 2 = 10
2   **a)** 30 pens
   b) 15 × 2 = 30
   c) (Yes)
Challenge 3
1

Number Problem	× or ÷	Number Sentence
One dog has 4 legs. How many legs do 2 dogs have?	×	2 × 4 = 8
Two identical ladybirds have 16 spots. How many spots does 1 ladybird have?	÷	16 ÷ 2 = 8
There are 9 boys in class 5. Class 6 has double the number of boys. How many boys are in class 6?	×	9 × 2 = 18
Ellie needs 40 straws for her party. They come in packs of 10. How many packs does Ellie need?	÷	40 ÷ 10 = 4
Joe saves 50p each week. How many weeks will he have to save to have £1?	÷ (or ×)	100 ÷ 50 = 2 or 50 × **2** = 100 100 ÷ **2** = 50

2   Appropriate number problem and answer.

Pages 52–53
Challenge 1
1   **a)** 4 boxes   20 ÷ 5 = 4
   b) 50 flowers   5 × 10 = 50
   c) 5 tables   30   10 × 3 = 30

Challenge 2
1 2 ice creams $2 \times 20 = 40$ 10p
2 30 fish $2 \times 15 = 30$ $30 \div 2 = 15$
3 25 snacks 50 4 days $20 \div 5 = 4$

Challenge 3
Examples:
1 a) 5 leaves each had 4 segments. How many segments were there altogether?
   b) Ben shared his 12 buns equally with Lizzie. How many did they each have?

Pages 54–55
Challenge 1
1 Half of each square shaded.
2 8 halves

Challenge 2
1 Each circle shows a clear division into halves.

Challenge 3
1 Each half is represented as an acceptable reflection and appears symmetrical.
2 Each object's missing half has been drawn and object seems complete.

Pages 56–57
Challenge 1
1 Each shape has two sections shaded to show one half.
2 Each shape has one section shaded to show one quarter.
3 Each shape has three sections shaded to show three quarters.

Challenge 2
1 a) $A = \frac{1}{4}$
      $B = \frac{3}{4}$
      $C = \frac{1}{2}$
      $D = 1$ whole pizza
      $E = 0$
   b) $A = \frac{3}{4}$
      $B = \frac{1}{4}$
      $C = \frac{1}{2}$
      $D = 0$
      $E = 1$

Challenge 3
1 a) 1
   b) 3
   c) 2
   d) 4

2

Whole	Halves	Quarters
1	2	4
2	4	8
4	8	16
3	6	12
6	12	24
10	20	40

Pages 58–59
Challenge 1
1 a) $\frac{1}{2}$
   b) $\frac{1}{2}$
   c) $\frac{1}{4}$
   d) $\frac{1}{2}$
   e) $\frac{3}{4}$
   f) $\frac{1}{4}$
2 0 $\frac{1}{4}$ $\frac{1}{2}$ $\frac{3}{4}$ 2

Challenge 2
1 a) 10 b) 5
   c) 15 d) 10
   e) 75 f) 25
2 a) < b) >
   c) = d) <

Challenge 3
1 a) 5 halves
   b) 13 quarters
   c) 22 quarters
   d) 41 quarters
   e) 25 halves
2

$1\frac{1}{4}$	$1\frac{1}{2}$	$1\frac{3}{4}$	2	$2\frac{1}{4}$	$2\frac{1}{2}$	$2\frac{3}{4}$	3	$3\frac{1}{4}$	$3\frac{1}{2}$

Pages 60–61
Challenge 1
1 a) Bag A
   b) 3 marbles
   c) 15 marbles
   d) 6 marbles
   e) $\frac{1}{2}$ OR half
2 Each pizza has one quarter coloured.

Challenge 2
1 a) Two items ringed.
   b) Four items ringed with a dotted line.
   c) Six items ringed with a zigzag line.

Answers

2 a) 12p
b) 6p
c) 18p

Challenge 3

Item	$\frac{1}{4}$	$\frac{1}{2}$	$\frac{3}{4}$
20 cm	**5 cm**	10 cm	**15 cm**
60p	**15p**	**30p**	**45p**
40 teddies	**10 teddies**	**20 teddies**	**30 teddies**
100 pencils	**25 pencils**	**50 pencils**	**75 pencils**
80 flowers	**20 flowers**	**40 flowers**	**60 flowers**

Pages 62–65
Progress Test 2

1 a) 15
b) 80
c) 21
d) 18
e) 50
f) 12

2 a) 8 tens, 7 ones
b) 5 tens, 9 ones
c) 4 tens, 1 one
d) 6 tens, 0 ones
e) 9 tens, 9 ones

3 a) 6
b) 3
c) 9
d) 12

4 a) 15   b) 12
c) 9   d) 25
e) 50   f) 30

5 a) 20   b) 40
c) 100   d) 90
e) 34   f) 24

6 a) Array shows 3 × 2 blocks shaded.
b) Array shows 6 × 2 blocks shaded.

7 a) 100
b) 500

8 a)

| 42 | 39 | 36 | 33 | 30 | 27 | 24 |

b)

| 45 |
| 40 |
| 35 |
| 30 |
| 25 |
| 20 |
| 15 |
| 10 |

9

38	43	48
46	51	56
24	29	34
28	33	38
82	87	92

10 a) >
b) =
c) <
d) <
e) =
f) >

11 a) 3 + 9 + 2 = 14
9 + 2 + 3 = 14 (or any other combination)
b) 4 + 8 + 6 = 18
6 + 8 + 4 = 18 (or any other combination)
c) 5 + 12 + 4 = 21
12 + 5 + 4 = 21 (or any other combination)
d) 11 + 7 + 1 = 19
7 + 11 + 1 = 19 (or any other combination)
e) 10 + 14 + 20 = 44
20 + 10 + 14 = 44 (or any other combination)

12 a) Group B
b) 6
c) 6
d) 8
e) $\frac{1}{4}$

Pages 66–67
Challenge 1

1 a) Snake D
b) Snake C
c) Longer
d) C, A, B, D
e) 51 cm

2 a) 4 m
b) 5 m
c) 2 m
d) 2 m

Challenge 2

1 a) 4 cm
b) 5 cm
c) 6 cm
d) 7 cm
e) 3 cm

Challenge 3

1 a) B and E
b) 2 cm or 3 cm
c) 6 cm or 7 cm
d) 11 to 14 cm
e) 4, 5 or 6 cm

Pages 68–69
Challenge 1
1. **a)** A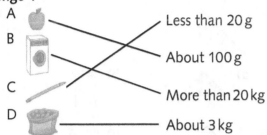
   Less than 20 g
   About 100 g
   More than 20 kg
   About 3 kg
   b) Pencil OR C
   c) Pencil, apple, potatoes, washing machine
   OR CADB

Challenge 2
1 **a)** 400 ml   **b)** 100 ml
  c) 500 ml   **d)** 300 ml
  e) 0 ml

Challenge 3
1 **a)** Scott
  b) Salima
  c) Less
  d) Rhona
  e) Scott, Rhona, Freddie, Salima.

Pages 70–71
Challenge 1
1 **a)** 128 cm   **b)** 44 cm
  c) Ben, Jack, Billy

Challenge 2
1 **a)** 10   **b)** 200 litres
  c) 20   **d)** 300 litres
  e) 30

Challenge 3
1 **a)** 600 g   **b)** 1000 g
  c) 1 kg   **d)** 1.6 kg
  e) 3.2 kg

Pages 72–73
Challenge 1
1 **a)** 20°C   **b)** 10°C
  c) 30°C   **d)** 15°C
  e) 35°C

Challenge 2
1 **a)**    **b)**

c)   **d)**

e)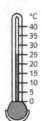

Challenge 3
1 **a)** 20°C   **b)** 5°C
  c) 0°C

Pages 74–75
Challenge 1
1 **a)** 3 o'clock   **b)** Half past 2
  c) 5 o'clock   **d)** Half past 7
  e) 9 o'clock

Challenge 2
1 **a)**   **b)**

c)   **d)**

e)

Challenge 3
1 **a)** 120 seconds   **b)** 48 hours
  c) 21 days   **d)** 104 weeks
  e) 120 months

Pages 76–77
Challenge 1
1   24p
2   45p   9 × 5p = 45p **OR** 5p + 5p + 5p + 5p +
         5p + 5p + 5p + 5p + 5p = 45p

3   Purse A: 51p
     Purse B: 93p
     Purse C: 83p

Answers

Challenge 2

1 Any five different combination of notes and coins totalling to £10

Challenge 3

1 a) £3
  b) £8
  c) £10.40
  d) £10.10
  e) £8.50

2 Any two combinations of notes and coins that total £7.84

Pages 78–79

Challenge 1

1 a) 40p
  b) 30p
  c) 60p
  d) 2 bunches
  e) 15p

Challenge 2

1 a) 50p
  b) £6
  c) £1
  d) £11.50
  e) £8.50

Challenge 3

1 a) 17p
  b) 28p
  c) 95p
  d) £4.10

Pages 80–81

Challenge 1

1

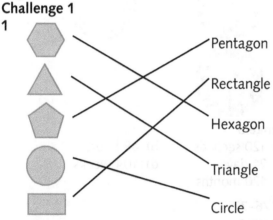

2 a) Pentagon
  b) Circle
  c) Rectangle (or appropriate alternative)
  d) Hexagon
  e) Triangle

Challenge 2

1 Any five different shapes that have properties of rectangles.

Challenge 3

1 Shapes B, C and F

2

3 a)   b)

  c) d)

Pages 82–83

Challenge 1

1

Shape	Has Six Faces	Has Circular Faces
Cuboid	✓	✗
Cylinder	✗	✓

2 a)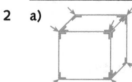

  b) 12 edges
  c) 8 corners

Challenge 2

1 Appropriate answers, e.g. four vertices and triangular base.

2 B and C

Challenge 3

1 a) Sphere
  b) Square-based pyramid
  c) Cone
  d) Cuboid
  e) Cylinder

Pages 84–85

Challenge 1

1 Any five different examples of shapes that have five straight sides.

Challenge 2

1 a) Hexagon
  b) Pentagon
  c) Square **OR** rectangle
  d) Hexagon

Challenge 3
1 Any six four-sided shapes.
2 Examples:
  a) Candle
  b) Washing machine
  c) Football
  d) Rubik's Cube

Pages 86–89
Progress Test 3
1 Any nine number facts that add up to 15.
2 £6.70
3 a) 16
  b) $4 \times 4 = 16$ **OR** $4 + 4 + 4 + 4 = 16$
  c) 40
4 a) 33 apples   b) 4 baskets
  c) 66 apples
5 a) 800 ml   b) 10 minutes
  c) 100 containers   d) 4 scoops
6 a) $1\frac{1}{2}$ b) 5 footballs
7 a) =   b) >
  c) =   d) =
8 a) 25 cm   b) 2 m
  c) 6 snakes
9

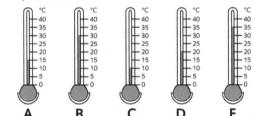

   A   B   C   D   E

10 a) 10p
   b) 50p
   c) £2.00

Pages 90–91
Challenge 1
1 Six arrows pointing down, left, up, right, down, left with alternate shading.
2 Ten arrows chosen show pattern of direction and colour.
Challenge 2
1 a) Objects repeated to form a recognisable pattern.
  b) Objects repeated to form a recognisable pattern.
Challenge 3
1 a) Recognisable pattern
  b) Recognisable pattern

Pages 92–93
Challenge 1
1 a)

20	20	19	19	**18**	**18**	17	17	16	16	15	15

  b)

1,1,1	3,3,3	5,5,5	**7,7,7**	**9,9,9**	11,11,11	13,13,13

  c)

5,5,10	6,6,12	**7,7,14**	**8,8,16**	**9,9,18**	10,10,20

  d)

100	100	90	90	**80**	**80**	70	70	60	60

Challenge 2
1 a)

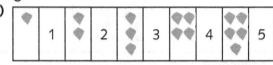

  b) 1, 1, 1, 3, 2, 2, 2, 6, **3**, **3**, **3**, 9, 4, 4, **4**, **12**
  c)

20	10	30	20	20	40	20	**30**	**50**	**20**	**40**	**60**

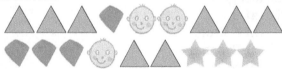

Challenge 3
1 Example. Shapes can change but quantities must not.

2 Any choice of shape that follows chosen number sequence.

Pages 94–95
Challenge 1
1 a) Z
  b) W
  c) X
  d) Y
Challenge 2
1 a) No
  b) B
  c) B
  d) A
  e) No
Challenge 3
1 a) Dog
  b) Mouse
  c) Mouse
  d) Snail
  e) Snake

Answers

Pages 96–97
Challenge 1
1 **a)** banana **b)** pizza
Challenge 2
1 Appropriate answers, e.g.

Challenge 3
1 **a)** Cat
   b) Dog
   c) Tree
   d) Castle

Pages 98–99
Challenge 1
1 **a)** 3
   b) Carrots
   c) Bananas
   d) 5
Challenge 2
1 **a)** 4 buses
   b) Cars   12
   c) 6 motorbikes
   d) 6 vans
Challenge 3
1 **a)** 8 snails
   b) 6
   c) A, B and D
   d) 12 snails

Pages 100–101
Challenge 1
1 **a)**

Animal	Tally	Total
Meerkat	卌 卌 I	11
Goat	卌 III	8
Duck	卌 卌 卌	15
Otter	IIII	4
Owl	卌 I	6

   b) Ducks **c)** Otters
   d) 2 **e)** 44

Challenge 2
1 **a)**

Hair Colour	Tally	Total
Blond	卌 II	7
Black	卌 I	6
Red	卌	5
Brown	卌 卌 I	11

   b) 5 **c)** Black and red **d)** 29
Challenge 3
1 **a)** and **b)**

Food	Tally	Total
Banana	IIII	4
Potato	卌 I	6
Apple	卌	5
Tomato	卌 卌	10

   c) 25

Pages 102–103
Challenge 1
1 **a)** 6 **b)** 9
   c) 2 **d)** Bella
   e) 16
Challenge 2
1 **a)** 5 **b)** 30
   c) Cats **d)** 20
   e) 60
Challenge 3
1

Cars	Monsters	Badges	Stickers

Pages 104–105
Challenge 1
1

Item	2 Wheels	No Wheels	More than 2 Wheels
A	✓	✗	✗
B	✗	✗	✓
C	✓	✗	✗
D	✗	✓	✗
E	✗	✗	✓

Challenge 2

1 a) Veggie Burger; 7
   b) 3
   c) Soup; Sandwiches
   d) Chicken and beans on toast

Challenge 3

1

Animal	Long Neck	4 Legs	Spots or Stripes
A	✓	✓	✓
B	✗	✓	✗
C	✗	✓	✗
D	✗	✗	✓
E	✗	✓	✓
F	✓	✓	✓

Pages 106–107
Challenge 1

1

Toy	Tally	Total
🧸	IIII II	7
🦔	IIII	4
⚽	IIII I	6
🤖	IIII II	7
🧒	IIII IIII I	11

Challenge 2

1 a) Dolls; 11
   b) 7 robots; Teddy bears
   c) 35 toys
   d) 3 aliens
   e) 17

Challenge 3

1
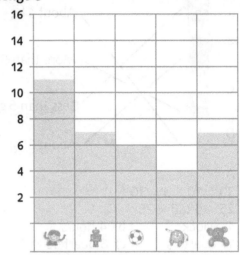

Pages 108–111
Progress Test 4

1 | 3 | 17 | 21 | 23 | 58 | 79 | 85 | 97 |

2 a) 30 + 20 = 50   50 – 30 = 20
      20 + 30 = 50   50 – 20 = 30
   b) 16 + 14 = 30   30 – 16 = 14
      14 + 16 = 30   30 – 14 = 16

3 a) 10 conkers
   b) Jay swims not as far as Hameed.
   c) 35
   d) 25

4

Leaf	Has Fewer Than 3 Points	Has More Than 3 Points	Smooth Edges	Jagged Edges
A	✗	✓	✓	✗
B	✓	✗	✓	✗
C	✗	✗	✓	✗
D	✗	✓	✗	✓
E	✗	✓	✓	✗

5 a) 11   11
   b) 9   9
   c) 15   15
   d) 50   50
   e) 30   30

Answers

6 a) A

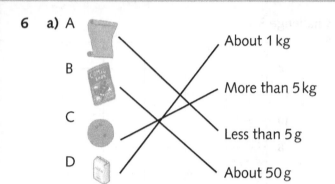

b) A

7 a) 30   10 × 3 = 30
   b) 25   5 × 5 = 25

8 a) 4 o'clock   **b)** 4 hours

9 a) 13 quarters   **c)** 6 quarters
   b) 11 quarters   **d)** 20 quarters

10

11 a)   **b)**

   c)

Progress Test Charts

Progress Test 1

Q	Topic	✓ or ✗	See page
1	Numbers and Counting		12
2	Using Two-Digit Numbers		30
3	Solving Number Problems		28
4	Doubling and Halving using Addition and Subtraction		26
5	Numbers and Counting		12
6	Numbers and Counting		12
7	Place Value		22
8	Counting More or Less		20
9	Numbers and Counting		12
10	Doubling and Halving using Addition and Subtraction		26
11	Doubling and Halving using Addition and Subtraction		26
12	Using Two-Digit Numbers		30
13	Doubling and Halving using Addition and Subtraction		26
14	Counting in Steps of 2, 3, 5 and 10		16
15	Solving Number Problems		28
16	Counting in Steps of 2, 3, 5 and 10		16

Progress Test 2

Q	Topic	✓ or ✗	See page
1	What is Multiplication?		38
2	Place Value		22
3	Finding Fractions of Larger Groups		60
4	Doubling and Halving using Multiplication and Division		48
5	Doubling and Halving using Multiplication and Division		48
6	What is Multiplication?		38
7	Solving Multiplication and Division Problems		50
8	Counting in Steps of 2, 3, 5 and 10		16
9	Counting More or Less		20
10	Less Than, Greater Than and Equal To		24
11	Solving Number Problems		28
12	Finding Fractions of Larger Groups		60

Progress Test Charts

Progress Test 3

Q	Topic	✓ or ✗	See page
1	Using Two-Digit Numbers		30
2	Standard Units of Money		76
3	Solving Multiplication and Division Problems		50
4	Doubling and Halving using Multiplication and Division		48
5	Measuring Weight and Volume		68
6	Fractions of Numbers		58
7	Fractions of Numbers		58
8	Measuring Length and Height		66
9	Measuring Temperature		72
10	Money Problems		78

Progress Test 4

Q	Topic	✓ or ✗	See page
1	Numbers and Counting		12
2	Using Two-Digit Numbers		30
3	Solving Number Problems		28
4	Tables		104
5	Doubling and Halving using Multiplication and Division		48
6	Comparing Measurements		70
7	What is Multiplication?		38
8	Measuring Time		74
9	What is a Fraction?		54
10	Quarter Turns and Half Turns		94
11	2-D Shapes		80

What am I doing well in? _____

_____

What do I need to improve? _____

_____